KU-008-792

THE FATE OF THE DINOSAURS

by the same author:

Noise Pollution
London's Drowning
Floodshock
Our Drowning World
Earth's Changing Climate

THE FATE OF THE DINOSAURS

New Perspectives in Evolution and Extinction

Antony Milne

Published in Great Britain 1991 by

PRISM PRESS
2 South Street
Bridport
Dorset DT6 3NQ

Distributed in the USA by
AVERY PUBLISHING GROUP
120 Old Broadway
Garden City Park
NY 11040

Published in Australia by
UNITY PRESS
Lindfield
NSW 2070

ISBN 1 85327 070 9

Typeset by Prism Press, Bridport, Dorset.
Printed by Biddles Ltd, Guildford, Surrey.

Contents

Extinction is no shame. It is, in one sense, the enabling force of the biosphere.

Stephen Jay Gould

My thanks go to Dr Angela Milner of the Natural History Museum, London, for pointing me in the right direction, and to Dominic Belfield, zoologist, and Michael Allaby, ecologist, for their help and advice.

Geological Eras

ERA	*PERIOD*	*MILLION YEARS AGO*
MESOZOIC	Cretaceous	65 140
	Jurassic	180 195
	Triassic	230
PALEOZOIC	Permian	280
	Carboniferous	345
	Devonian	395
	Silurian	435
	Ordovician	500
	Cambrian	570
ARCHAEO-ZOIC	Algonkian	2300
	Archaean	4500

Fig.1. The Geological Eras

Geological Eras

EARTH EVENTS	*BIOLOGICAL EVENTS*
Atlantic widens Widespread seas, new mountains	Disappearance of dinosaurs, ammonites, marine reptiles. First marsupials, placental mammals.
Pangaea breaks up First flowers appear	Widespread nature of dinosaurs. Appearance of first birds.
Ural mountains appear	First appearance of dinosaurs, paramammals, tortoises, turtles, marine reptiles.
Formation Pangaea, Appalachians and central European mountains.	Extinction of the trilobites and other invertebrates.
The forests expand	First reptile-mammals appear
Caledonian mountains appear	First terrestrial vertebrates, amphibians, sharks and armoured fish appear
First forests appear; also plants with seeds.	Appearance of first life on land above sea level; first invertebrates on land appear.
Mercynian mountains appear	First vertebrates, corals and cephalopods appear
Seas spread across the continents	First invertebrates appear: arthropods, molluscs and sponges.
Formation of the Earth	First organisms visible to naked eye.
Formation of the Earth	First signs of life.

AUTHOR'S PREFACE

Most authors like to believe that their new book on a well-worn subject fills a gap, and perhaps contributes some additional specialized knowledge that is in their own domain. I hope I have written just such a book. But I hope above all that it will appeal to those who love theories, ideas and speculation that is well rooted in mainstream scientific thought, and who like a book that covers a lot of up-to-date scientific ground.

The book's aim, therefore, is not to record just another 'Day of the Dinosaur', full of difficult to remember Greek names broken down into endless sub-orders and infra-orders; although as much as is necessary to know about dinosaur evolution is included. Nor does it detail any (often boring) narrative of the life, times and intrigues of eighteenth and nineteenth-century amateur fossil collectors; nor does it grind an axe in favour of warm-blooded over cold-blooded dinosaurs. Nor does it, in spite of its title, once more try to prove that the dinosaurs all died out when a giant missile struck the Earth.

It does, however, warm to the exciting new perspectives which some of these recent, and highly partisan, approaches have yielded, where argument and theory have been vigorously pursued in the scientific press, and attractively presented in more popular journals and in books, to the great benefit of knowledge and the advancement of science. Much credit must here go to the American academics whose names scatter the text of this book, and the universities and institutions that have supported them. Whatever one might think of the various new theories concerning the exact metabolic nature, form and structure of dinosaurs and why they became extinct, I hope the reader will agree with me that, at last, we are close to solving the mystery of their demise.

PART I

THE RISE OF THE DINOSAURS

Chapter 1
LIFE FROM SPACE?

THE 1980s have sewn radical new doubts about the nature of our solar system. There has been a revival of catastrophe theory, a return to doctrines over 200 years old, to explain extinctions. There has been much talk of comets, asteroids and exploding supernovae. There has been a greater emphasis on cosmic perspectives generally — such as the 'panspermia' theory (the notion that the 'seeds' of organic life drifted to Earth from outer space) — to explain both the origin of life on Earth and the frequent recurrence of plagues and viruses which have taken their toll of life in the past.[1] It is also now doubted that Pluto is really a planet at all, although, ironically, astronomers believe they have discovered a 'tenth planet' named Planet X. In addition some scientists believe the Sun, like many other stars in the galaxy, may be part of a binary system and have a twin.

But let us, for the moment, place these new theories within a longer historical timespan. It is natural that our understanding of the solar system and our place in it be subject to periodic revision, continuing with a process of learning and discovery that began long before even the invention of the telescope. Many of the theories may be taken as a kind of realignment of astronomical perspectives; a process of taxonomic refinement. Panspermia, for example, is a more rigorously defined statement that precursors of organic molecules are to be found in space, which in itself is neither an original nor a controversial finding.

However, recent spacecraft voyages have dissuaded astronomers from being too ready to categorize objects as satellites, asteroids, comets or even failed suns; or to neatly itemize inter-planetary debris, with planets being bigger than moons, asteroids bigger than meteors, and so on. In fact we now know that Jupiter, Uranus and Saturn have more satellites than was once detectable from Earth-bound telescopes, and some of the moons in the solar system are bigger than planets. Even some tiny red dwarf stars are hardly bigger than Uranus.

The idea that the planets were formed out of swirling discs of gaseous matter in space was first voiced by Emmanuel Kant more than two centuries ago. At the centre of Kant's disc was the Sun, and the planets were assembled from material that surrounded it. In the late eighteenth century the visionary French mathematical astronomer, the Marquis de Laplace, added some detail to Kant's vision by proposing that the planets emerged out of the rings of cosmic dust flung off by centrifugal force from the densest part of the cloud that was to become the Sun. These views were remarkably prescient for their time, since astronomers now believe that newly spinning protosuns do shed their outer rings of material in this way.

Deriving his theory from Newton's laws of angular momentum, Laplace postulated that a vast disc-shaped pall of cold gas was rotating in space in the vicinity of our present solar system. It gradually contracted to increase its rate of spin. We know that angular momentum can be expressed as the combined product of a rotating body's mass, velocity and radius, generally denoted as:

$$L = m\,v\,r$$

where L is the angular momentum, m is mass, v is velocity and r is radius. L, v and r are the variable factors, with a change in one being counterbalanced by a change in the other. Eventually, according to Laplace, v increased so much that centrifugal force at the edge of the nebula overcame the gravitational force exerted by the mass, and a ring of material broke away. Then, as the radius of remaining material shrank, other rings of material nearer the centre broke off. Some of Laplace's rings were thicker than others, thus attracting by gravity more of the material, to ultimately solidify into planets. The comets, meteors and other bodies were regarded as waste material.

The problem, however, was that not enough of the gaseous nebula could in fact have broken away from the embryonic sun, as later generations of astronomers were quick to point out. In fact, the planets possessed barely one per cent of the solar system's mass, but had somehow acquired 98 per cent of its angular momentum. The sun, in brief, spins some 100 times slower than it should do if the $L = m v r$ equation is correct.

Out of this difficulty, according to astronomer John Macvey, was born the celebrated 'other star' theory, which was to dominate astronomical thinking for about four decades.[2] Only an external event could have interfered with Newton's laws of celestial motion. Perhaps, it was surmised, another star had a near-miss with the sun, and supplied additional energy not accounted for in the angular momentum theory.

One famous theory was derived from the ideas of the popular astronomers Sir James Jeans in England, and T.C. Chamberlain in the United States. The approaching star pulled out by gravity great elongated filaments of hot material into a curved configuration, which then condensed into the planets. This had the effect of slowing the rate of rotation as the Sun was induced to emulate the filament around it. Perhaps, however, it was the Sun's twin star that suffered the collision with another star, leaving a large tendril of ripped-off gaseous material in the vicinity of the Sun which again would have a retarding effect on the Sun. Another theory suggests that a supernova explosion occurred, destroying the companion sun and scattering its material toward our Sun.

But it soon came to be realized that any breakaway

plasma would disperse into the void rather than condense into solid bodies. Rather, according to Lyman Spitzer of Princeton University writing in 1939, it would spread out around the Sun as a tenuous cloud. It could only have been relatively cool gases that could initiate accretion processes. The long-standing idea that the Earth started out as a molten mass, gradually cooled and shrank to its present size (a theory that still had adherents well into the 1950s), had to be abandoned.

The problem of the slowed rate of spin of the Sun and its large mass was solved when post-war scientists looked beyond the dynamics of what would happen to purely solid bodies functioning according to Newtonian laws. Once the central part of the nebula had undergone compression and nuclear reactions, the magnetic interaction between the hot, spinning mass of the plasma (the new sun) and a cloud of electrically charged particles would play an important role. So important, in fact, that such a cloud would have had a braking effect on the Sun's spin over aeons of time.

Fred Hoyle, in 1961, also suggested that the protosun had originally contracted by shedding more of its outer web, but managed to spread to the outermost periphery of the Sun, carrying with it the Sun's angular momentum.[3] It was not until 1972 that the idea of star and planet systems forming from other stellar material came into vogue, when Wilbur Brown of Wyoming University said that such systems were formed from the ejected shells of supernovae.

A word ought here to be said about supernovae, since they play a crucial role in the creation of planets and of life itself. Possibly, according to some scientists, they play a major role in destroying life (*see* Chapter 6). A great deal of the heavier particles that make up our bodies and the Earth itself are blasted out from incredible nuclear reactions taking place in dying stars (dying stars, in fact, both explode and implode). When most of the helium in a star is used up, the nuclei begin to fuse to become the heavier elements at the core, liberating new energy. With only 6000 years of life left the temperature rockets to 620 million degrees, to convert carbon into neon. When the core temperature reaches 1.3 billion degrees neon reacts to form oxygen. With just one year to go silicon is formed. Once iron is formed the limit of fusion is reached — no further energy can be liberated from it.

The core grows massively dense and gravity causes it to collapse in on itself at 25 per cent the speed of light. The core elements are smashed into sub-atomic particles, and huge quantities of neutrinos are blasted out by the shock wave at speeds of 11,000 miles per second, unleashing bright flares of light which astronomers can now, for the first time, observe. The supernova then becomes a black dwarf, a neutron star, or, if big enough, a black hole.

The region where the new solar system is to be formed is fortified by the particulate remains of the supernova, now charged with cosmic and X-rays, and other energy waves. Strong observational backing from scientists at the University of Arizona shows that a wave of gravitational energy sweeps through the gas clouds in the galaxy, with a mass at least one-thousand times that of our Sun. This causes the material to become very dense in parts, to rotate rapidly, and finally to collapse into smaller, opaque spheres as large as our solar system. Ultimately they contract further into molecular structures, into great dark swirling spheres that form the basis of young star-field systems. The cloud becomes very dense at the centre, leaving behind a dusty envelope that ultimately becomes a retinue of planets — like being thrown out onto a potter's wheel. And if our understanding of star formation is correct, then this means that most stars of the same size and density as our Sun will have planetary systems.

Soon the density becomes great enough for the spheres of gas and dust to collapse and start individual nuclear reactions. How does this happen? Gravitational forces alone, it seems, are not sufficient. Frictional heat plays a larger part — another example of the heat-energy theory at work. The cloud's rising internal heat can no longer be radiated away into space, and the dust cloud — originally resembling a giant gaseous ball like Jupiter — becomes a warm gassy *protostar.*

Astronomers can observe this happening because of the emission of infra-red energy, with dense clouds collapsing in on themselves at a rate of half a mile per second. The imploding discs of cosmic matter are rotating, and as the pressure increases the rotation speeds up like a flywheel, ultimately flinging matter out into space. This is one explanation why stars are often to be found in groups of twos or threes.

Material from the rest of the cloud globule continually falls into the embryonic sun, building up further mass, pressure and temperature, soon reaching thousands of degrees centigrade and beginning to glow like the T Tauri stars (glowing clouds of dense cosmic dust), which astronomers now think are the typical intermediate link between the dark globules and fully formed stars. As the sun shrinks it rotates faster, and flings off its outer, much cooler shell when it reaches escape velocity. These materials then flatten out as a disc surrounding the young star — to ultimately condense into planets. Some of them look and behave like the original gaseous protostar, but without the necessary mass to generate internal heat.

Astrophysicists are in dispute about the next stage. As the star is incandescently hot plasma (i.e. ionised gases with the nuclei of particles being transformed into the nuclei of other atoms because of the immense pressure of convection heat) a great deal of electromagnetic energy is created as escaping electrons and other particles are 'blown' into space, curving along magnetic lines of force. These particles rotate with the star and help to slow it, transferring the spin to the outer reaches of the globular dust cloud. This helps explain why most of the mass of the solar system is concentrated in the Sun, with the rotation spread among the planets; i.e. they have been 'disc-braked'.[4]

The Creation of the Planets

In the 1940s Otto J. Schmidt said that the planets were not created out of spun-off solar material but were 'captured' by the Sun from clouds of space dust. The theory, however, was weak, since it was unlikely such material could have solidified, in any manner that astronomers understand, into planets. It was in the Fifties, however, that Gerard P. Kuiper said turbulence in various parts of the cloud was likely to make the cloud sufficiently dense to flatten it out into a disc-shaped mass, with a large mass at the centre orbited by smaller ones.

Nowadays 'cold-welding' seems to be the most popular current explanation. The specks of dust consisting of heavier particles flung off by the Sun aggregate into larger lumps (a process which can be duplicated in a laboratory). Then a snowballing effect takes over, with the larger lumps able to grow larger still because their own gravity would attract other particles as they sweep along in their own orbit. Constantly colliding 'supergrains' would cohere increasingly under the redistributive influence of transferred angular momentum, aided by a gradually shrinking volume of space around the Sun. Soon an object reaches meteorite size, and then finishes up as a planet-sized object, all within a few thousand years.

We have already come across our first clue as to why species become extinct, and many others will emerge as the evolutionary story unfolds; all of the solar planets were formed by catastrophic processes. Earth not only grew by aggregation from a cloud of particles, but by collisions with other cosmic matter. It was the random collisions melding the rocky substances, plus turbulent accretions, that were to make up the inner planets. Some of the collisions were violent, but other planetesimals simply blended in with other collisions. (A planetesimal is a small, solid celestial body that may have existed at an early stage of the development of the solar system.)

Eventually supergrains of matter coalesced until the rocky fragments became large enough to attract more materials with gravity accreting at a faster and faster rate until they became full-sized planets. It was then that the electromagnetic forces became important, as they governed and determined the chemical compounds on Earth and the way they reacted with other elements. Radioactivity helped the solar wind to sift out the dense iron-nickel from the silicates, and was the main energy source in turning the inner cores of planets molten. Iron-nickel, with the silicates and non-metallic substances in gaseous form, soon rose to form the outer mantle of the Earth.

Then the largest planets themselves would act like mini suns, able to attract their own system of orbiting satellites. This reasoning alone casts doubt on the idea that the Earth's own moon was formed from its orbiting dust grains. The moon must have been captured later, although there is an interesting contemporary theory stating that the moon is the end-product of a tremendous collision between Earth and a giant planetesimal which reduced the careening object to pulverized grains. These then accreted into a moon by cold-welding techniques.

The matter of heat balances is important to the entire theme of this book (*see* Chapter 5). The constancy of heat both on Earth and in animal species has a bearing on life and death cycles. It was the level of heat which had a bearing on the chemical and gaseous make-up of the planets. Temperature determines the rate of condensation of gaseous matter into solids. In the early stages of cold-welding the planets must have been very similar in appearance, with massive outer gassy atmospheres. But it was the piercing electromagnetic field forces and the solar wind that blew away most of the atmospheres consisting of the very lightest gases from the planets nearest to the sun, to leave them with dense cores of rock and metal, the outer gaseous planets being less affected in this way (now we can see why the innermost planets of the solar system are solid, metallic and rocky, while the outer ones are icy and gaseous).

The proof that the planets were actually formed in this way arises from the fact that all of them have the same

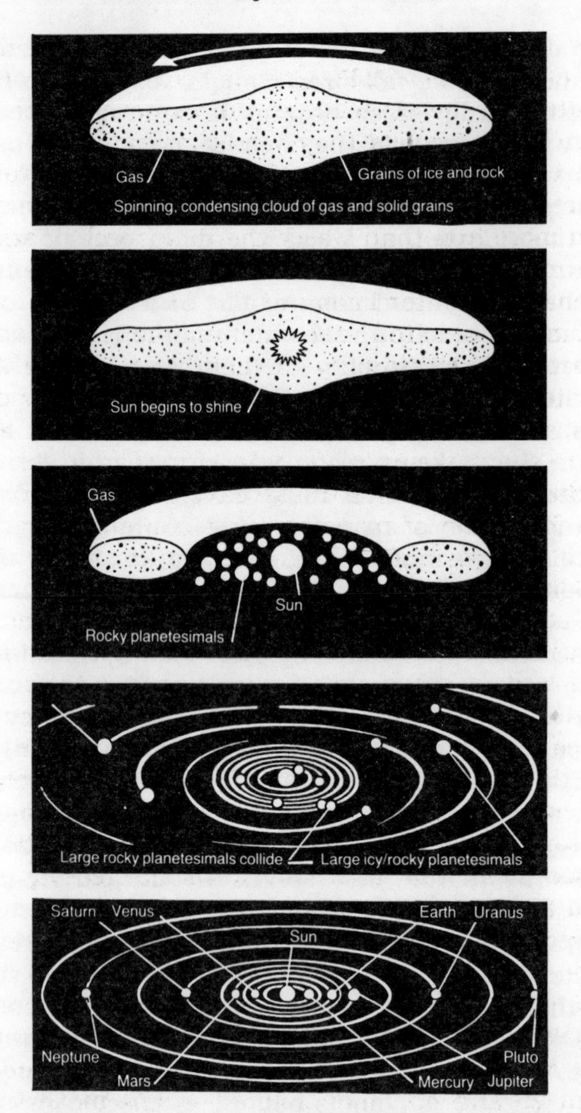

Fig.2. The formation of the planets. After the Sun was formed, piercing heat and radioactivity drove out the icy particles which formed giant balls of gas further out. The solid inner planets were formed from rocky grain cohering, and from collisions between planetesimals.

built-in mode of spin. Those with moons (which is most of them) orbit in a sun-like mode, with identical axial momentum. If the planets had simply been 'collected' by the Sun in its travels they would not all have similar orbital configurations.[5] In fact, recent spectroscopic measurements of meteorites suggest the Earth may contain more iron than was hitherto suspected, says Don Anderson of the Californian Institute of Technology in Pasadena. The outer layers of the Sun seem to contain forty times more iron and calcium than carbonaceous chondrites. Traditionally it has been thought that meteorites were formed at the same time as the solar planets, and the composition of them was used as a 'fossil guide' to the make-up of the planets. Now, it seems, the meteorites may not have the same origins as the Earth.[6]

The formation of planetary systems and the size and chemical composition of each member are, then, remarkably regular: given a sun-sized star a computer generally comes up with a similar range of planets to that in the solar system, with small rocky planets closest to the star and the large gaseous ones further out (due to the effects of gravity, orbits would later space themselves out along the lines of our own solar system). So the ordered structure of the Sun, with its retinue of planets, repeated what was occurring a million times around other similar stars of the same mass and density — what is known as 'main sequence' stars. The Sun is such a star, creating enough warmth to make a planet like Earth habitable.

However, getting the temperature just right for life to arise on a planet is so unlikely that the chances of it occurring are billions to one against. Both Mercury and Mars have gravitational fields too low to retain an atmosphere. Mercury's distance from the Sun, varying from between 29 and 43 million miles, would make the Sun seem three times bigger in the Mercurian sky than it would on Earth. This means that its surface temperatures must be literally scorching, frequently reaching 700°C. Temperatures during the long Mercurian night would probably plummet to below zero. Its axis rotation time is equivalent to 58.5 Earth days, and there would be no atmosphere to retain the high 'daytime' temperature, nor shield the planet from harmful cosmic rays.

Venus, however, is afflicted with a lethally hot and corrosive atmosphere. Although some 30 million miles further away from the Sun than Mercury, its heat-retaining surface gases of carbon, oxygen and sulphur make the surface temperatures some 80°C hotter. Because it is closer to the Sun than Earth, Venus probably contained less water than Earth. When the interior began to heat up as a result of radioactivity, volcanoes and fumeroles erupted, pumping out vast quantities of carbon dioxide. In addition, the fierce solar heat released further carbon from its surface rocks. The heat was retained by the celebrated 'greenhouse effect', making the atmosphere hotter and very much denser, which in turn released further CO^2 from the rocks — the classic vicious circle.[7]

Mars, on the other hand, suffers from not having a 'greenhouse' problem. Mars is too far away from the Sun, and is too small. This means that any water vapour would turn to ice, because its low gravity would only retain a tenuous atmosphere with very little carbon dioxide to help trap the much-reduced solar warmth. Not only that, but continuous 124 mph winds surging around the planet whip up the ochre soil high into the atmosphere, turning the sky pink, virtually obscuring the already wan sunlight.

Earth was lucky. Its distance from the Sun was just right to sustain life, providing the right kind of equable temperatures and flowing water. Earth's warmth, like Venus's, also came from its interior, as heat-producing radioactive minerals decayed at the mantle, reaching nearly 5000°C at the molten centre. The core, indeed, was probably molten even at its protoplanet stage, and this formed an intensive magnetic field.

All biological life thus adapted its metabolism to the level of solar warmth, and became adjusted to the strength of solar radiation prevailing at the time. We must remember, too, how beneficial are Earth's seasons. All of the planets circle around the Sun elliptically in the same direction as the Sun itself rotates. But individually they orbit at different speeds depending on their distance from the Sun, the innermost rocky planets orbiting much faster than the outer balls of frozen gaseous vapour, such as Jupiter, Saturn and Uranus. And they do not all rotate in the same mode. Most of them are tilted in their orbits at an identical angle to Earth's (the exception being the inner and outer two planets), proving that they have all evolved together.[8]

The planetary seasons greatly determine temperatures and climatic patterns. The Venusian 'year', as determined by the *Mariner 2* satellite probe launched in 1962, turned out to be some 225 terrestrial days, although its rotation was so slow (243 Earth days) that its 'day' is actually longer than its 'year'. The later *Mariner 10* probe, launched in 1974, confirmed that the dense sulphuric clouds moved across the sky so rapidly as to create violent storms at the surface. The Martian year is nearly twice as long as Earth's, although its 'day' is virtually the same.

The density of Earth is partly explained because of its metallic core which is responsible for the magnetic field. This field, again, is important, since without it, as we shall see, great harm to living creatures could occur as a result of what goes on in outer space. With a density of 5.52 grams per cubic centimetre, Earth is more solid, in a sense, than the other three terrestrial planets, as well as being the largest (its diameter is 8000 miles).

Many theorists believe that Jupiter could well be a 'failed star', not quite big enough to trigger nuclear reactions. However, if Jupiter were slightly bigger or had a rather less regular orbit, scientists believe the other solar planets would follow far less predictable orbits.

The theory of planetary formation is not perfect, and possibly a part of the sequence is missing or misunderstood. But it reflects the astronomical consensus of what probably did happen. Kuiper's theory, though, implies that the Sun and planets were created simultaneously, but modern theory says the star had at least reached its proto stage before being robbed of its momentum when filaments of hot gases were thrown out. Even now the nebula theory is being challenged by a new book on the origin of the planets, by two British astronomers. Michael Wolfson and John Dormond say, in effect, that Sir James Jeans' 'capture' theory may be the right one after all. They say that a close encounter between the Sun and a more diffuse star-like object in the cluster tore off a hot filament of gases. This cooled and condensed into the solar planets, which were later themselves subject to more distortion by the Sun. The inner solid planets, it is claimed, were created by a catastrophic collision between two giant planets orbiting where Mars and the asteroid belt now reside.[9]

The Creation of a Liveable World

In Britain in the late 1980s the idea of the Anthropic Principle, a hitherto obscure doctrine in physics, became the subject of conversation at cocktail parties, was probed exhaustively in a 700-page book and was summarized in popular science magazines and television documentaries. The Anthropic Principle has stunning implications for virtually all branches of science. It decrees that intelligent life in some way *selects out* its own actual universe from a variety of possible alternatives. The biochemical evolution of the Earth, including the creation of the reptiles, dinosaurs and the mammals, has as its sole purpose the ultimate evolution of intelligent beings capable of understanding the laws of physics, and thus able to speculate about the origin and density of the universe.[10]

In its extreme form, as enunciated by Brandon Carter, a cosmologist now at the Paris Observatory, it says that the conditions we observe in the universe must include the various electrical and gravitational constants that hold all planetary matter together and thus give rise to intelligent terrestrial life. If they did not, we would not be here to observe them. One curious conclusion is that the dinosaurs *had* to die to make way for humans: this is the starkest meaning of the Anthropic Principle.

Of course, the argument is highly teleological and incapable of scientific proof, but it has been rigorously defended by brilliant minds and cannot be lightly dismissed. It received a new boost in 1990 when John Gribbin and Martin Rees published their *The Stuff of the Universe*, where it was asserted that the universe came into existence solely in order to create a carbon-based intelligent life-form on just *one* planet — Earth. This is because the conditions needed for the Big Bang had to be so finely tuned — to an amazing one part in 10^{60} (a big number even on a cosmic scale) — that the most minute variations would not have permitted any life at all to develop.[11] And according to physicist Tony Roth the Principle could well be true because many earlier broad cosmological theories have paved the way for later testable discoveries, as has the Copernican Principle, and even the Principle of Beauty which bore great fruit in the theory of electroweak interaction, which recently united the electromagnetic and weak nuclear forces.*[12] In any event, the theory of environmental selection by variant genotypes within an evolving species is quite common in biology.[13]

There are, incidentally, several opposing theories to the Anthropic Principle. Human logic says that if in fact we are all made from the ashes of long-dead stars, then everything which now *is* should be ultimately related to what first *was*. There could also be parts of the universe

* The Copernican Principle, derived from Nicholas Copernicus' sixteenth-century calculations proving that Earth was part of a solar system orbiting the Sun, says that Earth is no more privileged than any other celestial object. The Principle of Beauty says that the most symmetrical, or most beautiful, theory frequently offers the best insight into scientific problems.

containing more anti-matter than matter. It is possible that there is an undiscovered and unique set of equations that describes all the forces and particles in the universe in a kind of Theory of Everything that would do away with the Anthropic Principle.

But there is another way in which the Anthropic Principle, or more strictly a Biothropic Principle, could be indisputably self-evident, if it means that animal and biological life on just one planet has its genesis in cosmic forces. Certainly Earth needs its Sun and moon, and neither could exist without the Milky Way first coming into being, and so on. But even if the question of necessity is ruled out, the notion that events in outer space influence Earth and its atmosphere, as well as the nature and distribution of animal life, is now a substantive truth. In every case only one well-known galactic force is involved — gravity.

Gravitational influences between and within bodies in space are detected as perturbations and periodicities which, over millions of years, can actually mould planetary surfaces. Some years ago great controversy was stirred up when two astrophysicists published a book, based largely on Chinese research, suggesting that solar fluctuations triggered geophysical disturbances here on Earth.[14] Such a reaction was curious, as Santa Barbara biogeologist Preston Cloud reminds us, because the very rocks beneath our feet rise and fall like the ocean tides, but imperceptibly, in a twice-daily rhythm in response to the gravitational pull of the Sun and moon.[15] From time to time Earth suffers bombardment from meteors and comets, and meteoritic impact has been a major planet-shaper and climate-modeller. Further, prolonged rain storms on Earth have been observed to occur in line with certain alignments of the moon, and perhaps of other objects in the solar system.[16]

The important thing to remember is that climate is influenced by events above and below the Earth. The Earth can generate much more internal heat than has hitherto been suspected. For this reason the golfball picture of the planet, with regular internal layers and smooth, uniform demarcation zones, has fallen out of favour as it has encouraged what are probably quite wrong estimates of temperatures at the core. And it only partially explains how deep internal forces drive the movements of the Earth's crust, and why volcanoes and earthquakes occur.

The volume of the ocean basins has fluctuated over time and this has a great influence over virtually every other life-generating and sustaining force ever discovered. It is now clear to Earth scientists that the extent of ocean water and its depth and shallowness, and the way it periodically invades the land and then retreats, has been a major climatic determinant. The arrangement of the oceans in relation to the continents plays a decisive role in creating and sustaining life on Earth. So it is not only Earth's distance from the Sun that is important for living things, but the way the solid parts of the Earth are spread across the globe.

From this geophysical fact other organic facts follow. The main chemical ingredients for life are derived from combinations of carbon, hydrogen, oxygen and nitrogen, which we can abbreviate to CHON. These atoms, as we have seen, are spewed out from dying stars; but it would take at least a billion years for a star to use up all its helium energy and explode, hence life could not exist in a universe consisting of the shortest-living stars, since the necessary elements such as carbon would not exist. The fundamental CHON atoms would combine and recombine to form organic compounds of greater and greater complexity. They formed gases and ultimately amino acids, the so-called building blocks of life. Later the changing composition of atmospheric gases, as a result of new forms of microbial life emerging out of the amino acid stage, also affected the climate.

Biologists Lynn Margulis and Dorian Sagan are supporters of the well-known Gaian hypothesis (see over). In a recent book they pointed out that the laws of physics dictate that the sun's luminosity must have increased by

30 per cent during the past four million years. And yet, they say, Earth's temperature has remained stable, implying that life has been regulating the composition of gases, and the biota must have been controlling its own environment. The great shake-out of particles at the beginning of the solar system's life, the way they can recombine to give altered secondary atmospheres, is a feature of great importance in explaining how life took hold. The bonding of atoms via their electrons plus the temperature they function at, gives a clue as to how molecules form. Another clue is tidal forces and the atmosphere. All of the original planets had extensive atmospheres of hydrogen, carbon dioxide and helium which was swept away by the intense solar radiation. In Earth's case it developed a secondary atmosphere through organic evolution, say supporters of the Gaian hypothesis.[17]

The Gaian hypothesis was first promoted by the atmospheric chemist James Lovelock, and has stimulated a great debate about whether the fundamental controls on the Earth's warming gases are biological or abiotic (physical).[18] The Gaians suggest that the micro-organisms such as plankton produce vast quantities of dimethyl sulphide (DMS). When DMS interacts with atmospheric oxygen it can create clouds over the ocean and can determine how much solar heat reaches the surface.[19] More heating of the Earth either by solar or gaseous factors will increase the amount of plankton, which will in turn lower temperatures. This is known as a 'negative feedback' response, where additional inputs into the equation have a regulatory effect. (A 'positive feed-back' is where a situation is exacerbated by the inputs.)

But the abiotic school of thought has equally persuasive arguments at its disposal. Life on Earth, the biosphere and the hydrosphere, including the dynamics of climate, are a densely woven story of the carbon atom. Without carbon there would be no organic chemistry, and no life. Scientists interpreting the greenhouse effect today know that the oceans act as a massive 'sink' containing fifty times as much carbon as the atmosphere, mostly dissolved in the form of bicarbonate. Seafloor minerals and continental sediments contain some 100,000 times as much carbon as the atmosphere.[20]

Whether carbon dioxide is regulated by organic life or the physical Earth itself is admittedly one of the central battlegrounds in the Gaia debate. But it is worth remembering that early in Earth's history the continents were much smaller than they are now. This fact alone explains the Sun/climate anomaly. The Sun must have been considerably dimmer and cooler shortly after the Earth was formed because thermonuclear processes had not yet reached their peak. So Earth should have been commensurately cooler. But the geological record shows no trace of any cooling until much later. One reason could be the fact that the absorption spectra of stars show the rare and chemically inactive gases such as krypton, neon and xenon to be much more common than here on Earth. This probably means that Earth may initially have been hotter than some scientists suggest, since even these heavy noble gases were driven off when the hot gaseous molecules reached high escape velocities.

More probably it was the abundant presence of carbon. According to geophysicist James Walker of the University of Michigan the continental carbon was divided up among the atmosphere, the primeval sea floor and the oceans themselves.[21] But as the continents grew they gleaned more of the carbon from the oceans and atmosphere. As carbon dioxide retains infrared heat better than any other gas, and as the temperature began to fall, it follows that the early Earth must have had an atmosphere predominantly of CO^2, and that atmospheric CO^2 must have diminished.

At this stage, however, we can perceive a curious twist to the argument, with a blending of the Gaian and abiotic schools. The world of four billion years ago was swathed in carbon dioxide, producing a massive Venus-like greenhouse effect. Life, in fact, would have been impossible for most higher forms, with only the hardiest microbes

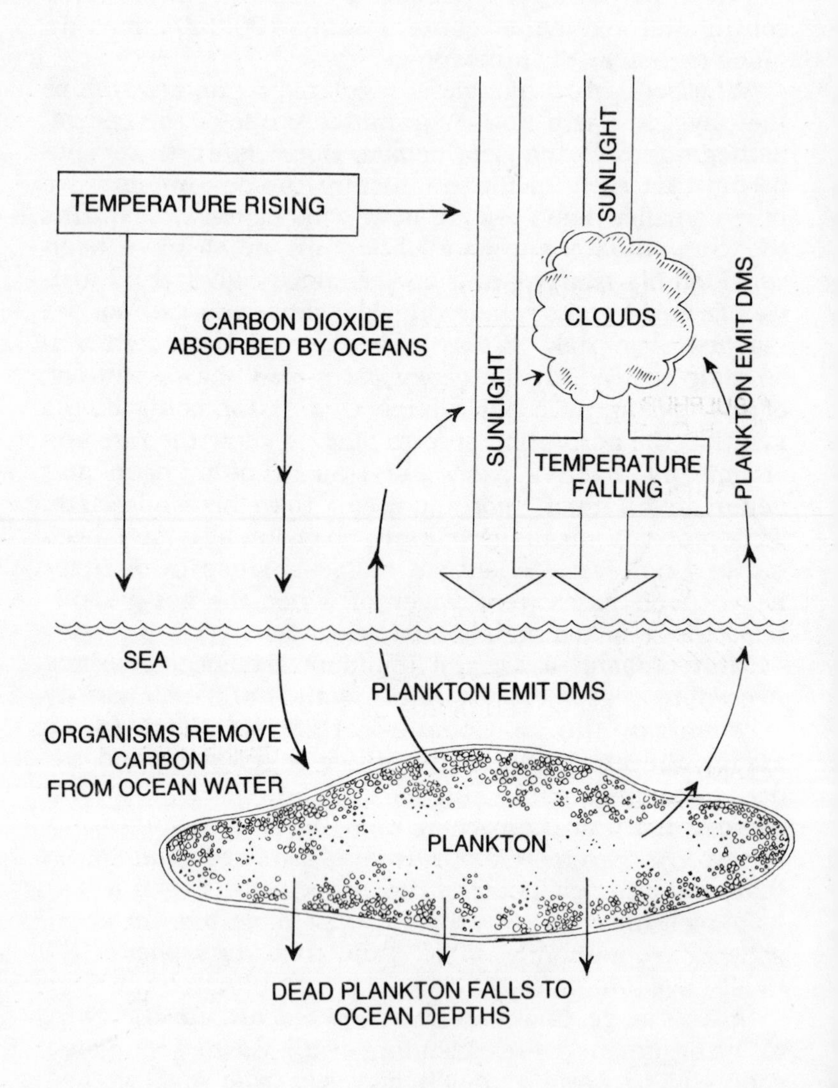

Fig.3. The Gaia school of Earth temperature regulation.

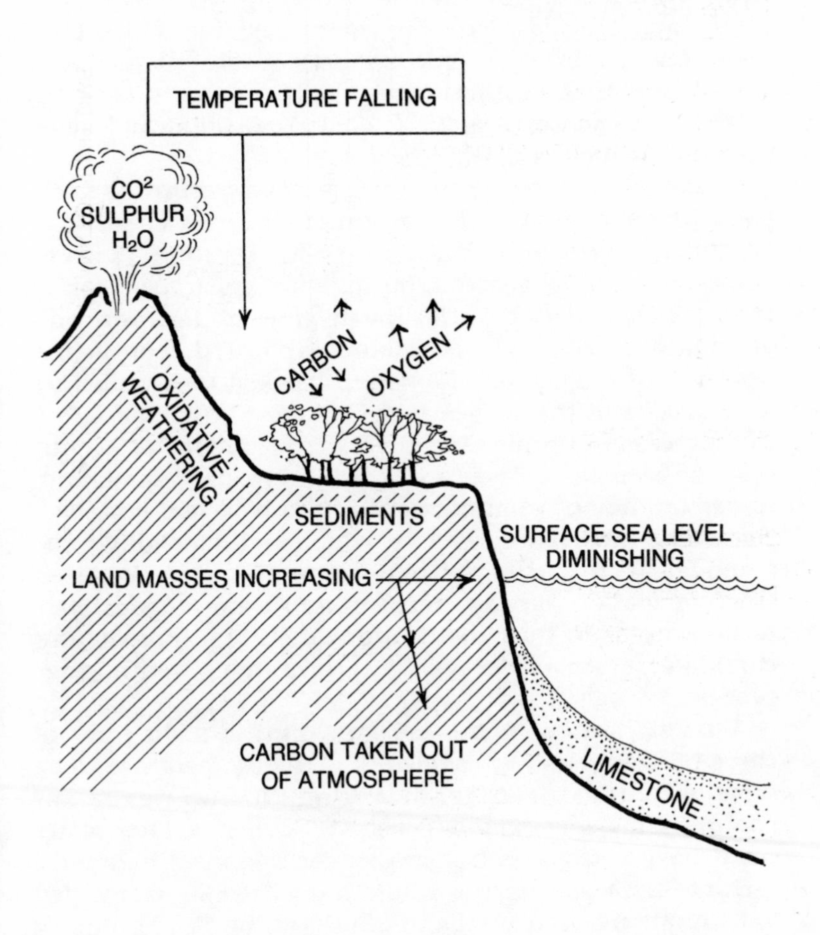

Fig.4. The abiotic school of Earth temperature regulation.

surviving. But in a recent *Nature* article, David Schwartzman of Howard University, and Tyler Volk of New York University, maintain that primitive microbes staved off the greenhouse effect, so helping weathering (where more carbon gets buried in surface soil) by stabilizing the soil against erosion. A stable soil soaks up water, and allows acid to break up bedrock and create more soil. Rocks can also be broken up by microbes finding their way into cracks and crevices, exposing more surfaces which can then absorb more CO^2, and so forth. Schwartzman and Volk reckon that weathering was helped along to such an extent that the Earth is 35°C cooler than it would have been on the lifeless Earth.[22]

There is another way in which we can confirm the role played by climate and the carbon atom in the creation of organic life. We know that shortly after creation a violent spate of seething geotectonic activity must have taken place as the interior of the Earth grew hotter, new outgassing occurred and new islands appeared. Continents began to drift apart as the molten basaltic lava welled up from cracks in the seabed (*see* Chapter 8). But the scientific process of unravelling the exact sequence of events has not been easy. Even now there are many gaps in our understanding of what happened. It is believed, however, that some early igneous rocks, which had to solidify from a molten state, are proof that some atmospheric gases (such as carbon dioxide, nitrogen and water vapour) had to be present to turn them into something approaching earthlike rocks and clays by a primitive 'weathering' process.[23]

Flowing water, evident from younger sediments of three billion years ago, shows that the temperature must have been above freezing, indicating the Sun was not as cool as some scientists suggest.[24] Sooner or later water must have acted as a transporter for dissolved materials. Much of the water vapour would have been supplemented with ammonia and methane, and possibly, according to the evidence of rocks and the sedimentary record rich in silica, a little nitrogen and water vapour existed, but oxides of carbon predominated. The existence of silica suggests that ammonia was not an important gas, as it would have left much more limestone and dolomites.

The fact that our atmosphere and oceans are largely the product of outgassing by volcanoes is important. Volcanoes bring both life and death to Earth. They are erratically distributed around the planet, and the gases they emit vary in outgassed particular matter. However, when scientists first tried to work out how life on Earth got started, they had to perfect their understanding of the original primeval gases, and whether and how such gases changed their nature. One early assumption was that the first atmosphere on Earth consisted of gases rich in nitrogen and hydrogen, such as ammonia and methane. What is crucial to Earth's history, however, was that, unlike any other member of the solar system, it became a wet planet. One of the most important compounds is water, since water acts as the suspension medium for the vital chemical reactions which later sustained living systems. The mobile and metamorphosing earthly environment was responsible for the nature of early life on Earth, allowing different mixes of chemicals and minerals to blend under changing conditions.

We can be reasonably sure that the first water vapours were the natural result of condensation of the outgassed products of volcanoes. Most volcanoes emit around 60 per cent water vapour, 24 per cent CO^2 and 10 per cent sulphur, with the rest a varied mix of chlorine, nitrogen and hydrogen. The saturation resulted in rain, which fell for a very long time. A shallow sea began to emerge. Then the water vapour which condensed into oceans must have come from early volcanic activity and from the gaseous components of the solid Earth itself. Whether the outgassing of vapour occurred continuously or episodically is still a matter of scientific dispute. According to W.W. Rubey, formerly of the US Geological Survey, the rate of sea formation was probably faster at the beginning phase of volcanism and then tapered off.[25]

Later scientists thought that a succession of 'heat

waves' arising from radioactive decay in subsurface materials, and by the gravitational collapse of the solid lithosphere, was responsible for sweating out most of the water vapour. Later still the 'heat waves' were thought to be produced by catastrophism — recent calculations by Tokajumi Matsui and Yutaka Abe of the University of Tokyo suggest that billions of tons of meteorites crashed into Earth to drive off the volatile materials and to turn the surface into seas of magma.[26] Nowadays it is thought that not much new water is generated by volcanoes or by hot springs and vapours, which speeded up in time through land-ocean changes and continental drift; the vast majority of this is today believed to be recycled hydrospheric water.

Then, while the vapour condensed into oceans the high percentage of CO^2 would have caused a runaway greenhouse effect. After about 800 million years the mean surface temperature probably reached a maximum of 300°C, and the pressure half as much higher than today. The Earth and the oceans absorb much of the carbon compounds, much of which combines with calcium and magnesium to become limestone, which holds much of the CO^2. The Earth, in fact, has enormous 'sinks' for both oxygen and carbon, and it is when these become full that excesses accumulate in the atmosphere. Occasionally, however, shortly after organic life got a toehold, Earth's surface water may have evaporated by frequent asteroid impacts, which had the effect of sterilizing large areas of the globe, according to a group of American geophysicists based in California.[27] This would explain why single-celled organisms appeared 3500 years ago, but complex creatures such as mammals and reptiles did not appear until 500 million years ago.

James Kasting, of the Pennsylvania State University, also told the American Association for the Advancement of Science in February 1990 that, judging from ancient craters on the moon, collisions with asteroids nearly 300 miles in diameter, hitting the Earth at more than 100,000 mph, were much more common in the early history of the solar system than today. A collision would vaporize the oceans and raise atmospheric temperatures to 3000°F. It would take months for the rocks to cool off, and about a thousand years for the steam in the air to condense back into the seas. Only primitive life-forms in deep ocean ridges would have survived, the descendants of which survive today as plants and huge worms that live around sulphurous vents in the ocean floor.[28]

Life From a Chemical Soup?

But how *did* primitive life-forms first appear on Earth? Scientists are still not certain about this in spite of what the reader might believe. The more established theory is that the rise and fall of water levels, the advance and retreat of lakes and the resultant slimy muds, were also catalysts, or concentrators, bringing new chemicals within sediments into contact with other organic chemicals. Land-based organisms carry their watery heritage with them, locked inside their cells.

It is possible, of course, to conceive of alternative biochemistries and other liquid states. The *Pioneer* and *Voyager* probes of the 1970s hinted that Jupiter probably had no solid surface at all. Its atmosphere takes up the outermost 600 miles of its 38,000 mile radius, and is mostly hydrogen, with some methane (CH_4), ethene (C_2H_2) and ammonia (NH_3), plus water vapour (H_2O) and phosphine (Ph_3). It is also possible that two of these constituents in their liquid form substitute for water as a suspension medium for organic compounds.

However, the evolution of complex living creatures (not just humans) would be based on a high information content greater than could be found, say, in the bacteria of the atmosphere of the Jovian planets. Spontaneous natural selection requires an advanced level of organization at the micro-biological stage. It also needs long periods of ecological stability during which evolutionary epochs can bring about the necessary organic synthesis.[29] But little is known about the chemical 'soup' of the oceans, and therefore the type of life that first became selected and concentrated there and which would have been different from that occurring, say, in rivers or stagnant pools. Such would have a tendency to dry out or cool according to changing climatic conditions. Did the amino acids and nucleotides come about spontaneously in the warm shallow waters of a young Earth? Was there a natural organic synthesis?

Possibly not. For one of the most important discoveries of the twentieth century was the fact that some molecules in space actually appeared to be similar to those that make up amino acids. This was suspected because when the particles collide they emit radio signals. Each molecule has its own distinctive wavelength, enabling them to be identified. In 1965 one cyanide (CN) was discovered in space in this way, as well as the hydroxyl molecule (a primitive water vapour element with one atom of hydrogen missing). Later, ammonia was found in a cloud near the centre of our galaxy. Formaldehyde molecules (consisting of four atoms: 2 of hydrogen, 1 of carbon, 1 of oxygen) were also found by scientists at the National Radio Astronomy Observatory in West Virginia. To date about one hundred organic molecules necessary for the evolution of life have been detected.

The veteran astrophysicist Sir Fred Hoyle, with his colleague Chandra Wickramasinghe, believes that these molecules were probably brought to the surface of the Earth via the tiny dust grains on meteorites and comets, or on interstellar winds, and succeeded in germinating life in the fertile conditions of Earth at the time.[30] Lynn Margulis and Dorian Sagan maintain, however, that neither the chemical soup explanation nor the 'panspermia' theory is necessary.[31] Because of the fact that organic life is built up from CHON (they mention others, phosphorous and sulphur), the natural energy and gravity cycles of the Earth can by themselves make the molecule bonding and the building blocks of life.

There are certain clues to what could have happened. Water itself is, of course, a combined source of oxygen just waiting to be liberated via the breakdown of water vapour through photosynthesis. All this helps oxygen accumulate, and carbon compounds 'sink' into rocks where they can recombine with oxygen. But there is the problem of oxygen 'sinks' existing as well, which could use up the oxygen produced by recombining with any of the CHON molecules. Carbon atoms can form bonds not only with themselves but with the atoms of important atmospheric gases, oxygen and nitrogen. The outer electron of the carbon atom can combine with four hydrogen atoms to form a carbohydrate; oxides would be produced, preventing free oxygen from accumulating.

One conventional understanding of Earth's primitive atmosphere, based on the famous Urey-Miller laboratory experiments of the 1950s, was that primitive gases such as hydrogen, ammonia, methane and water vapour, periodically energized by flashes of lightning, gave birth to the strange new combinations of CHON. These then formed the molecules known as the amino acids and nucleotides. The life-generating experiment has now, after varying the mix of gases, been repeated many times, and each time the building blocks of life have been easily enough manufactured. Moreover, the conditions have also been varied, with equal success.

The suggestion that catastrophic heating itself could be the prime mover gained in credibility. Could shock-waves have brought about organic life? Or, for that matter, meteoritic impacts? The Urey-Miller experiment was repeated by Israeli researchers Askiba and Burit Bar-nun at Cornell University, by subjecting ammonia, ethane and water vapour to a shock wave. This event momentarily raised the temperature inside the flask by several thousand degrees, leaving behind four new amino acids.[3 2]

There is now an extensive literature reporting on the various life-generating experiments, backed with much speculation and theory. The panspermia hypothesis does not deny that life started in a 'primeval soup', it merely shows that life can come together in a variety of different ways.[33] One recent theory says that the common ancestor of all living things was a bacteria that consumed sulphur compounds, and lived in a volcanically hot environment that was hot enough to boil water. James A. Lake, writing in *Nature*, says that similar bacteria survive in hot volcanic springs or around vents or fissures in the deep ocean floor.[34]

What can be stated with some certainty is that Earth's unique combination of water and heat started the whole process off, as all the laboratory experiments would appear to confirm. A consensus seems to have emerged that some kinds of blue-green algae started to extract hydrogen from the planet's richest resource, the oceans. The H_2O bonds were much stronger than for any other hydrogen compound, but yet the algae managed somehow to develop a secondary photolytic reaction centre for re-absorbing high-energy light that could split the water molecules up. Because they had so much more energy available to them, cyanobacteria exploded into a vast array of different forms.

At last the road to more complex creatures — an amazing event in the history of the universe — began to reveal itself. Soon amino acids linked together in groups and chains of ever-increasing complexity, when the small molecules became large ones, such as proteins, nucleic acids and enzymes (i.e. all those we regard as being the real precursors of living matter). The new bacterial forms were versatile and energetic, and could engulf other bacteria. Life therefore became a form of chemical interaction, with replicators able to change gradually through the mechanics of evolution.

Life, in a sense, spread across the globe by a form of networking. Exactly how this happened is intriguing. A process known as thermal polymerization probably linked amino acids to form proteins (the basis of the plastics industry is to use polymerization to assemble related molecules into chains called polymers). Scientists at the Institute of Molecular Evolution at the University of Miami believe that the creation of life may have been quite rapid by evolutionary standards, simplicity quite quickly becoming complexity. A protocell probably came into being a few hours after the formation of proteins. Likewise natural organisms can create polypeptides, which are small polymers made of amino acids. As enzymes they speed up the chemical reactions of life. These polypeptides in turn form into extremely long chains of proteins. An experiment was once performed where the acids were heated in a dry state to 150°F, to be immersed in warm water later. They then took on the appearance of microspheres, behaving and looking like living cells.

Early life on Earth may have had many false starts — as many as ten — with each based on a different biochemistry. Scientists now believe life is based on an infinite variety of polypeptides, with a great potential for unrealized polypeptides that could be used in living systems. Cells grouping together in colonies meant that they could be replaced when they died, with the group as a whole surviving intact should any membrane wall of any individual member be accidentally punctured. Soon living colonies became the norm, and they began to proliferate.

The Importance of Light

Using the hydrogen atoms in ocean water with the aid of light energy was a remarkable evolutionary event, but there were still abundant geochemical problems to overcome. One by-product of photolysis (photolysis is the dissociation of water vapour through ultraviolet radiation) was that once oxygen began to appear it soon combined with whatever ammonia or methane was then present, and built itself into rock structures. Being very reactive it combined easily with minerals in the Earth's crust. At this stage no green vegetation existed, so only photolysis could have created oxygen. This occurs at great heights where atmospheric densities are very low, so that many of the liberated hydrogen atoms escape into space before colliding with oxygen atoms and becoming water molecules. The remaining oxygen atoms simply accumulate in the atmosphere.

Furthermore the process of splitting H_2O (and CO_2, for that matter) is easily reversed. Carbohydrates can recombine with oxygen to form CO_2; and hydrogen, once O_2 is present in the atmosphere, can recombine with oxygen to form water vapour. It is also clear that the non-biological breakdown of water vapour in the outer atmosphere would not by itself be enough to get life started. For until such time as oxygen in sufficient quantities could create an ozone shield, organisms subject to the lethal wavelengths of ultraviolet radiation could not exist on land except in the shade, or in shallow water within reach of suitable wavelengths of light for photosythensis to take place.

Let us look for a moment at the ozone shield. Ultraviolet (UV) radiation between 100 and 200 nannometers (that is, in the UV-C range) is absorbed in our present atmosphere by O_2 in the upper layers. This process splits the molecules into two single atoms, most of which recombine to form O_2, especially at night in the absence of solar radiation. Some, however, form a tri-atomic form of oxygen, O_3, called ozone, which strongly absorbs UV radiation at wavelengths between 200 nm (nanometre: a millionth part of a millimetre) and 300 nm in the UV-B and UV-C region. Then again, simply by doing this, it decomposes again into O_2 and O. Not only that, but oxygen can react with ozone to form two oxygen molecules (O and O_3 = $2O_2$). Ozone is also decomposed through other chemical reactions.

Through all these processes, then, ozone is continually being created and decomposed. We know, in addition, that the ozone layer, between 15 and 35 km above the Earth's surface, removes all UV-C radiation. But the layer itself is astonishingly thin. If forced down to form a layer near the Earth's surface it would only be five millimetres thick. Nevertheless the existence of the ozone layer implies that it is almost inevitable that early organic life — algae — must have got started in murky waters, below which the rays, especially UV-C, with a powerful intensity overlapping with the X-ray region, could not penetrate.

However, ultraviolet radiation must also have aided evolution. Although it caused mutations — frequently devastating and biocidic (causing the death of life, as in biocide) — on very rare occasions it spurred evolution. Those early genera that survived had a built-in immunity to ultraviolet radiation, thus helping in the struggle for continued survival. The evolution of the biosphere — indeed the very existence of the biosphere — came about by taming the worst excesses of incoming solar radiation by screening out the more harmful rays.*

As the algae became more complex some form of repair mechanism probably occurred, especially for those existing in shallow watery environments (*see* Chapter 6). In the meantime, oxygen could not have accumulated at this early stage (without biological help) because it would absorb the same ultraviolet rays. So equilibrium would be maintained with oxygen recombination being continually counter-balanced by water splitting. In any event oxygen-breathing organisms could soak up the excess gas as soon as it was produced. We have also seen that there were 'sinks' in the form of sedimentary rocks. So early oxygen levels would have been periodically depressed, in order to allow the first photosynthesizers to carry on evolving.

Nevertheless vegetation soon started to proliferate — because liverworts, mosses, ferns, coniferous plants and flowers appeared in quick succession, introducing carbon dioxide into the atmosphere in ever-increasing quantities. This was achieved through photosynthesis: plants were

* The fact that there are some micro-organisms in laboratory experiments which seem to be immune to very high and potentially lethal doses of radiation is one reason why Fred Hoyle insists that organic life must have existed on the surface of meteorites or other airless bodies, and would have survived unadulterated ultraviolet radiation before entering Earth's atmosphere.[3 5]

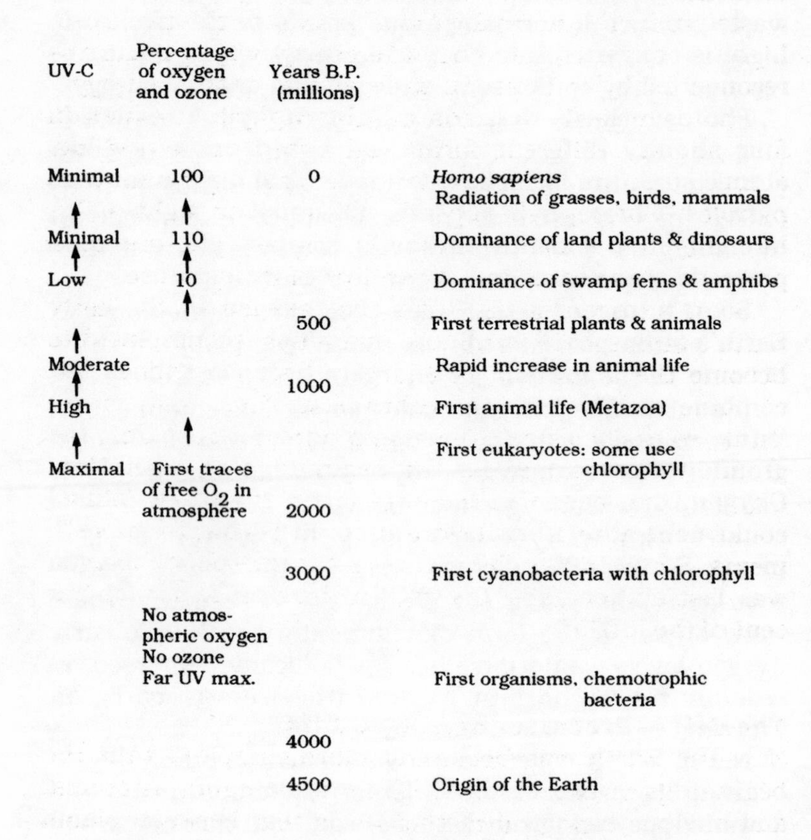

Fig.5. How the threat to organic life from ultraviolet radiation has decreased over millions of years, as molecular and tri-atomic oxygen has increased. (Adapted from New Scientist*)*

able, by converting sunlight into energy, to live solely on sunshine, air and water, combined with other simple elements gleaned from the soil. They produce oxygen as a waste product, a normally lethal poison to the plant cell. Light is converted into chemical energy which in turn is reconverted by combustion or decay into organic energy.

Photosynthesis depends on chlorophyll. It exists in four slightly different forms, all based on a ring-like atomic structure formed by four chemical units known as pyrroles. Chlorophyll depends naturally on light, CO_2, humidity and warmth, although primary photoreaction proceeds at the same rate regardless of temperature.

Soon nitrogen, a trace gas that existed in the early Earth's atmosphere in minute quantities, accumulated to become the dominant gas, largely because it does not combine readily with the other basic four, so it is not 'sunk' so easily. Once oxygen and nitrogen appeared, the groundwork was laid for the beginnings of vegetation. Oxygen, too, began to increase faster than new 'sinks' could neutralize it, or before it could be buried in sediments. Some 1800 million years ago atmospheric oxygen was fast approaching the present level of about 21 per cent of the total.

The Cell — Precursor of Animal Life

Now the Earth was becoming more complex, with the beginnings of two kinds of life-form emerging. A crucial and obvious biological distinction is that between plants and animals, but a more important distinction is the way life is based on one type of cell or another. Needless to say, plants had to evolve before animals: i.e. they had to become the food for which primitive animals had in turn to colonize the land to seek. Vegetation indirectly created the protective ozone shield by pumping oxygen into the atmosphere. More importantly, it held the soil in place, helped alleviate flooding and moderated the atmospheric temperature.

Evolution soon enabled some cells to take in, as food,

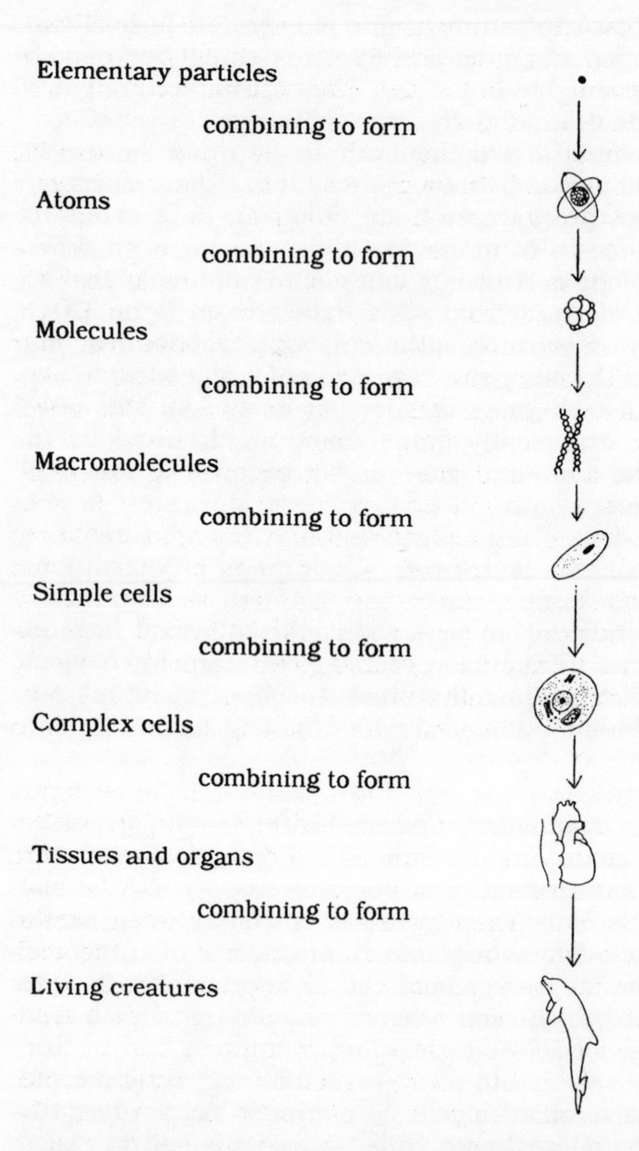

Fig.6. Evolution as a ladder of progression, where the tiny and simple aggregate together over time to become the large and complex. (Adapted from Peter Russell: The Awakening Earth*)*

highly organized matter, such as other life-forms. Thus ready-formed vitamins and proteins could be absorbed without them having to be laboriously manufactured from gases, other minerals, or light energy.

Both plants and animals are made up of *eukaryotic* cells, although the first simple cell was the *procaryotic* type. Most procaryotes, using oxygen, depend on the simpler process of fermentation for their energy, where chemical food is broken down and burned inside the cell. This cell, derived from the earliest life-form on Earth, although microscopically tiny (an earthworm, for example, contains more than a billion cells) is still a very complicated chemical factory. Bacteria and blue-green algae are chemically very similar, both being of the procaryotic kind, and are often known as cyanobacteria. Cyanobacteria still exist today, like the purple sulphur bacteria which glean hydrogen from hydrogen-sulphide to provide sulphur compounds as oxidation products. Even higher organisms resort to fermentation as an auxiliary process when the oxygen supply is insufficient. Throughout prehistory they have formed great carpet-like layers, the so-called stromatolites, like those found in Zimbabwe nearly three billion years ago, slowly calcifying into limestone.

Nevertheless in the early days many of the other forms of bacteria died off in vast numbers. Some, the precursors of fungi, could only survive in the dark. Ultimately the demand for nutrients became more than could be supplied by the Sun. The way out of this predicament was to alter their own environment through their own metabolism. Some micro-organisms began to manufacture their own nourishment and at the same time provide a food-chain base for all other emerging creatures.

Soon a new strain of oxygen-mediating species arose, enabling evolution to proceed as fast as carbon could be buried in the sediment. The procaryotic cell began to change its form somewhere between 1.3 and 2 billion years ago, when it evolved a kind of brain, developing a small nucleus at the centre that could control the entire chemical manufacturing process. It thus became the eukaryotic cell. Food could be turned into energy via respiration, in which glucose is either derived from food or from photosynthesis. The cell developed a new way of consuming oxygen to make carbon dioxide and water to release energy.

We know that the eukaryotic cell is much larger and more efficient than the procaryote, carrying more genetic material. Both cells exhibit the same biochemical pathways in photosynthesis, hinting at a common origin. Some scientists suggest that the procaryote is a degenerate eukaryote, one that has lost its unique features. The new eukaryotic cell was porous, and surrounded by a membrane a few ten-millionths of a millimetre thick. It trapped inside the cell the building blocks of raw materials such as amino acids which were converted into larger molecules. The cell also housed a watery fluid known as the cytoplasm, which contains the organelles that control a sequence of chemical processes converting food to energy. A certain type of organelle is the powerhouse, and is called the mitochondria, converting food into energy. Another set of ingredients, known as the ribosomes, are responsible for making new protein molecules. More importantly the larger molecules manufactured inside the cell were later to be called the dna.

Today every single living thing that has ever lived, from a bacterium to a plant or a fully formed animal, has been built according to specifications laid down in the molecules of the dna called chromosomes. The specific components of chromosomes in turn are called genes, that familiar word describing how detailed information about the whole organism is passed on from generation to generation. In fact, dna acts as the chemical basis for heredity, serving as a template for its own replication and for instructing the sequence of amino acids and proteins.

The dna molecule is composed of the string of compounds called nucleotides, having four bases attached to them: A C G T. What is surprising about dna is that A and T will cling together, and so will C and G. A single dna chain, immersed in a pool of nucleotides, will latch on to its counterpart and thus be able to duplicate itself when the double chain is split apart. The dna message is copied out onto the rna strand, which is almost identical to the dna (and which actually generates it). This rna strand is called ribosome, and is somewhat like a computer tape which the rna can zip through to specify the correct amino acid.

The early eukaryotic cell soon evolved an efficient method of sustaining its own growth, for the properties of a particular cell (such as its shape and its abilities) are determined by the chemical processes that go on within it. These chemical reactions would normally proceed very slowly were it not for Nature's invention of complex molecules called enzymes, which we have also met before in this chapter. Enzymes are a form of protein made out of a very precise order of amino acids. Particular molecules will attach themselves to the enzyme surface so that atoms can be transferred rapidly. The enzymes, in effect, determine the function of all multicellular organisms. Possibly the chromosomes carry a 'blueprint' of all the enzymes in the cell, guiding their manufactures.[36] In a human being one cell contains 23 pairs of chromosomes. Each time a cell divides, the chromosomes also divide.

A word of caution. The concept of an advantageous mutation spurring evolution is the standard model with which evolutionists still work. But this view has been challenged in a recent issue of *Genetics* by Barry Hall of Rochester University, following on from research done in 1988 by John Cairns, a respected molecular biologist. Hall says that mutations in the genetic code of some bacteria occur more often when they are *useful* than when they are not. Both types of mutation — harmful or useful — may arise at different rates depending on stresses inherent in the environment. When, for example, colonies of bacterial species in a laboratory were deprived of a certain kind of amino acid needed for growth, they produced mutant strains capable of producing their own amino acids.[37] The implication from Hall's research is that organisms can adapt their genes to suit their environment, and that the production of genetic mutations is not necessarily separate from natural selection. In turn it means that adaptive evolution may be faster than biologists had thought.

It is with this startling new perspective in mind that we view the eukaryotic cell's rapid multiplication as it spread throughout the world's oceans, finally replacing all the free-floating molecule strands that preceded it. One great improvement involved the outer membrane of the cell. Some became sticky and could clump together in colonies. In this way the individual cells could protect themselves from danger. Different cells took on different functions. A division of labour of newly forming sense organs arose. Some cells, for instance, became more efficient at assimilating food, and became the gut of new organisms. Others acquired a harder outer membrane to later become skin-like materials.[38] Other cells especially sensitive to light became vestigial eyes, and those susceptible to vibrations became hearing organs, and so on. These new improved organisms were much better able to survive and avoid danger, to find food and to reproduce, multiplying ever more rapidly. They became yet more complex, true progenitors of real plants and animals.

But by now even more fundamental differences were beginning to appear. In the case of plants the organic energy is derived from a primary source, and in the case of animals and carnivores, consuming both plants and meat, it is secondary. In other words mammals and living creatures generate kinetic energy and heat when food is burnt in respiration. Even the bacteria that feed on animal and plant wastes derive their energy from the Sun. The ecosystem as a result is shaped and controlled through the action of absorbed light energy. Furthermore, plants utilize carbon dioxide in the atmosphere that animals eject, and animals glean yet more energy from the environment by utilizing the oxygen that plants eject by converting it into new combinations of chemicals.

It is this complicated self-regulatory process that has

made such a profound impression on the Gaianists; i.e. the notion that living matter itself both defines and maintains the conditions necessary for survival, and that the chief physical features of the Earth are biologically controlled. Furthermore this 'biothropic' principle, as we referred to it earlier, implies that Man is the ultimate beneficiary of this chain of events. With his imagination stirred Man could consciously 'terraform' other planets to bring about similar Earth-like conditions. Carl Sagan, the distinguished astrophysicist, suggested in 1961 that Venus could be made habitable to earthlings if it were seeded with blue-green algae, which would split up the carbon and oxygen molecules to glean the carbon necessary for the glucose and carbohydrate diets they would need. Similarly Michael Allaby and James Lovelock suggest that living organisms could be introduced into the sterile environment of Mars to warm up the atmosphere.[39] The carbon thus liberated would stimulate a mini greenhouse effect very quickly indeed (owing to the incredible speed at which micro-organisms multiply) and bring the temperature up to around 80°F in the northern hemisphere.

In the meantime, on Earth, we have reached the stage in our narrative where simple unicellular creatures have become multicellular creatures, the earliest known of which would have been the jellyfish.[40] Jellyfish-remains dating from the Archaeozoic/Cambrian boundary have been found, for example, in Australia. Many of the other early multicelled creatures were similar to jellyfish and, like sea anemones, circular in appearance. They were simple creatures, made of two layers of cells, one on the outside and one on the inside, with every part of the body within reach of the outside for the efficient transport of nutrients and wastes. Flatworms and tapeworms succeeded in inventing a third layer of cells in between the other two. Blood vessels in this new middle layer gave genuine and unique scope for developing internal complexity by linking up inner and outer layers.

Most of the main animal groups appeared quite suddenly 600 million years ago in what has become known as the 'Cambrian explosion'. According to Peter Cook and John Shergold, of the Bureau of Mineral Resources in Canberra, this event probably occurred through a sudden release of phosphates from the deep ocean brought up by a change in ocean currents or crustal movements.[41] Contrasting designs in worms occurred. Many could successfully survive in a harsh environment, and their bodies enabled them to evolve into larger and more diverse creatures. Ultimately they turned into giant flies, octopuses and tadpole-like animals. Others, like sponges, consisted of a colony of cells with a porous skeleton. However, James Valentine of the University of California suggests worms may have evolved from arthropods (jointed-legged creatures) rather than the other way around, because they had cavities in their middle layer (known as coeloms) which they used as skeletons to support themselves when burrowing. These coeloms have since atrophied in insects.[42]

The earlier unicellular species were by now well adapted to their environment, and only changed their biochemical functions because of the genetic 'errors', the mutations, that periodically occurred. But all the while these mutations, as well as the natural hazards of the environment — the temperature variations, the strength of solar radiation, the salt concentrations in ocean water, for example — all acted to diversify populations. They grew in numbers and complexity. Soon complex life-forms emerged, based upon chemical arrangements that differed from each other. Then other living organisms provided them with further opportunities, and through symbiosis (a mutually-dependent state of living together) they created, finally, animals.

Chapter 2
THE FIRST CREATURES

THE FIRST creatures emerged on Earth as a result of biological processes becoming ever more complex and, some would argue, ever more mysterious. Fortunately in trying to fathom out what happened next we have the advantage of the known laws of science. For example, we know that a biological system of animals should function like any other system. There must be some flow of energy and information, and some dynamic mechanism to use this information. The system must also have a structure such as an ordered set of channels through which the energy can be directed.[1]

What the neo-Darwinists have in their favour is almost irrefutable evidence arising from recent advances in the techniques of molecular biology and genetic engineering which virtually prove that modern species did evolve from common ancestors.[2] However, for a variety of reasons animals become less similar as the evolutionary process continues and molecule patterns begin to differ. Populations that belong to a specific group with common ancestry are known to have a single genetic system.[3]

The dna determines the activities of the cell and thus the organism's shape, form and function. Together the sum of all genes possessed by the individuals in a population is known as the *gene pool*, and copies of the genes survive in other bodies to be replicated. Furthermore the information flows can be changed over time as a result of mutation in the genetic code or by the individual's own genetic system, and some information might thus be eliminated. These changes in genetic information are passed on to offspring, a process responsible for what the neo-Darwinists call natural selection.

Hence the individual soon knows, from this genetic information, what its form and structure is (its morphology), and it begins to acquire behavioural characteristics. The blueprint for the adult creature, derived from both parents of any species, is the *phenotype*, and it is the phenotype which most biologists think is the genetic mechanism for evolution. This does not mean that a 'blueprint' in the engineering sense exists. The idea that acquired characteristics or congenital injuries (or calluses on the knees of camels, for example) are passed on from father to offspring, was once taught by the Lamarckian school and is now discredited. The genetic code, in fact, merely 'tells' the organisms how to assemble their particular haemoglobin and bodily chemistry. Organisms throughout history have become increasingly efficient, with 'selection' driving this genetic process forward.

Once self-replication had got under way molecules could pile on other more complex molecules in a loose association. If the less successful variants were not multiplying quickly enough an element of competition would also mean that poor replicators will be less efficient at acquiring the means of survival to keep the new species going. Eventually they reached a stage where they became integrated units. One theory says biochemical mechanisms to control the regulation of growth of some part of the organisms, say hormones or fur, may themselves become faulty, or the conveyed messages over aeons may change, or the genes may suddenly switch off to inhibit the growth of winter fur of an animal used to northern climes. If this is so, it could very likely be the scenario for extinction as a result of genetic errors or failures. In other words, it was when replication went wrong, when random errors occurred in the copying process, that the process of evolution and extinction forged ahead, to populate the world with offspring slightly different from their progenitors while other genera died out.

Here we ought to stress that evolution is not spurred simplistically on by 'random mutation', as the Oxford zoologist Richard Dawkins points out in his book *The Blind Watchmaker*. He is in particular concerned to rebut the argument of anti-Darwinists who claim that if mutations are merely 'random' they cannot be biased towards improvement. This is not to deny that random changes are not automatically beneficial, but the direction towards improvement comes from natural selection; i.e. from genetic changes that were an aid to survival. This is where biogeology comes more clearly into the picture. For instance, Dawkins refers to other kinds of non-random influences, such as the 'mutagens' that can cause cancers: for example, X-rays, cosmic rays, various chemical and radioactive substances.[4] Some part of an organism's chromosomes have a high rate of turnover of genes, and are called 'hot spots' which locally have a very high mutation rate. If mutagens like cosmic rays are present then all normal mutation rates are boosted.

Darwinism implies that some species must triumph over others not because they are fitter in some genetic or athletic sense, but because they are better able to survive in a competitive ambiance. In other words, they are 'better designed' not simply by *chance* (which would be the case if a large single-step mutation occurred), but by the accumulation of myriads of single-step mutations aided, over long periods of time, by the natural weeding-out process of those species less successfully adapted for long-term survival.

The idea of fitness is a subtle one, and depends greatly on genetic make-up rather than morphology. Definite physical events in the universe and on Earth act as a non-random thrust towards improvement. Amelioration occurs in some groups, but most organic change centres around a 'set of basic designs' then reflecting a saga of accumulating excellence.

Our view is that competition between species drives

life onward, with one form gaining space in the crowded ecology of the Earth only by forcing out others. However, it is probably true to say that Darwin's theory does not dominate biological thought as much as it used to. Natural selection is important, but may not apply universally.[5] As we have seen, some genetic change may not take place at random, or events may occur more quickly than Darwin believed. Another view holds that evolution depends heavily on environmental change,[6] so the struggle is often more with the wider world than with other species. Evolution requires very many difficult and coordinated genetic modifications, which clearly are coterminous with, or causally related to, geophysical changes.

How Animal Groups are Sub-divided

As time passed, scientists felt they had to organize and categorize the growing number of different phenotypes and their evolving descendants. They came up with a taxonomic hierarchy: phyla, class, order, sub-orders, genus (or genera) and species. Phyla can be divided into orders, orders into families, families into genera, which in turn are sub-divided into species. Genera, sub-species, etc., can succeed each other temporarily or permanently, and if permanently either abruptly or gradually by transformation.

The lion and tiger species are to be found within the genus *Felis*. Humans are members of the *phylum* vertebrata, *class* mammalia, *order* primates, *sub-order* Anthropoidai, *genus* Homo, and *species* sapiens.[7] However, no new phyla have appeared since the Cambrian period, some 500 million years ago, hinting, as Gordon Rattray Taylor once pointed out, that perhaps evolution has actually come to a halt.[8]

It is interesting to note that there have, so far as we know, only been thirty-one phyla in all of pre-history; nine have become totally extinct. Up to twenty vanished basic phyla of soft-bodied creatures have been identified from the fossil record of 520 million years ago — all utterly different from anything known today. Recent discoveries by Cambridge University paleontologists have revealed that the prehistoric creatures that roamed the primeval oceans were even stranger and more varied than we can imagine. Some had five eyes and mouths like elephants' trunks, and yet others appeared to be walking on stilt-like appendages, and had heads with no features at all, not even mouths.[9]

Curiously, one pre-eminent group of creatures that included not only the dinosaurs but birds and humans — a group that has lasted to the present day — was the bipedal or two-legged walking variety that had a group of sense organs at one end. An even larger phyla would be the vertebrates, to include not only all animals but everything with a skeleton, articulated limbs or a shell, such as crabs or insects.

It is clear that these larger groups or phylas have arisen because of what can only be called *co-evolution*: the changes to take place in their body types have been more or less continuously in tune with evolutionary changes in their habitat or environment. For example, the running animals with long legs abide in areas with no thick vegetative cover in which to hide. It has been pointed out that the modern ostrich and the Cretaceous *Struthiomimus* lived in the same sort of exposed habitat, and thus evolved similar bipedal running gaits.[10] Likewise a long neck gives a good vantage point to warn of advancing predators. In other words genetic and geographic forces operate in tandem, as a form of biogeology or biogeography.

A further example is the modern Australian lungfish, of the genus *dipnoi*, which has a substance in its brain that lowers its metabolism when rivers seasonally dry up, as if it were hibernating,[11] so it could survive while others died off. But this also means that as the morphology of species matured over great periods of historic time, other forms became more adapted to life on land, while yet others returned to the water to escape the increasing competition.

It would be a mistake to make too many sweeping generalizations. It is still not easy to come to terms with the inexplicably uneven impact of the environment on changing animal morphology. For this reason some scientists argue that changes in the environment are not a satisfactory explanation because they are not sufficiently unique, and whatever phenotypic modifications occurred might just as easily be renewed or reversed.[1 2] The eminent evolutionary biologist from Harvard, Stephen Jay Gould, points to the weird ocean-going creatures mentioned above, where entire communities of soft-bodied organisms were preserved intact in a freak mudslide. The Burgess Shales, as they have become known, is proof to Gould that chance is more important than natural selection, since they all died off without issue whereas some minor offshoots of the extinct phenotypes, like the Crustaceans, went on to become important stepping-stones on the evolutionary ladder from which mammals finally emerged. This suggests that human beings arrived on Earth more or less by accident, since no special feature marked off the Crustaceans as being 'fitter' in a Darwinian sense.[1 3]

But even a random mudslide extinction serves to reinforce the fact that evolution involves the survival of individuals via the interchange of energy among contemporaneous plants and animal organisms, and between the Earth itself. Furthermore each living thing, plant and animal reacts and competes with other members of its species. The first-order consumers (the herbivores) ate the plants, and the second-order consumers (the carnivores or meat-eaters) ate the herbivores. Populations divided into predators and prey, or hosts and parasites. This interrelatedness was akin to a web rather than a chain — with complicating overlapping functions — since a great many organisms were parasites on all sorts of organic and biological life. But evolution ploughed on remorselessly, enabling only the most adaptable to go on to the next stage. This relationship, for very long periods, was in equilibrium. It was when it was disturbed — when the predator-prey ratio, or climatic changes, occurred — that events were set in train which would lead to inevitable extinction.

A major difficulty arises when fossil species disappear for good as the physical environment, over millions of years, inevitably changes. Herein lies an important clue, but also a further puzzle. When we try to ascertain whether climate and geophysical turbulence could have affected evolution, we have to ask whether the changes in environment killed off species or whether there was simply a gap in the record. Was there a sudden appearance of a new species, or were there cases of gradual succession? The way ahead for paleontologists trying to build up a knowledge of evolution has been spurred on by the revolution in plate tectonics, which has provided a better understanding of the stratigraphic record and fossilized data. For example, the presence of fossil salt lakes suggests hot and arid conditions; dune sand and soils cemented with minerals indicate deserts.[14]

The stratigraphic record can at best link these two important phenomena — the emergence of species and the long-term atmospheric temperature changes. In other words, the stress forces of the Earth, in particular stresses brought about by severe changes in climate, and the changing chemical nature of the surface, may have contributed to the great expansion, and occasional contraction, of species within narrow phyla that evolved from the recognizable solid life in the oceans (oceanic life started, probably, in the Algonkian period, somewhere between 670 million and 600 million years ago).

Stress can manifest itself in more than one way. For example, Conway Morris of Cambridge University's Earth Science Department believes that the comparative emptiness of the early oceans indicates minimal inter-group rivalry, but that 520 million years ago the competition began to hot up, and the basic animal groups, evidenced from parts of the fossil record of Northern America and Australia, were reduced from 55 to around 35.[15]

Another kind of stress is of a more abstract nature. Physicist Paul Davies talks about the progression to more complex organisms via the evolutionary ladder, and the steady accumulation of complexity which is time-assymetric; it defines an arrow of time from past to future.[16] The geophysical is also time-assymetric, of course. For example, the Variscan orogenic cycles started in the Paleozoic era were said to be responsible for the mountain chains of North America, Europe and Africa. This period of geophysical upheaval lasted until the late Carboniferous and early Permian times and was responsible for modifying Earth temperatures in the process.

Time-assymetry thus implies that while the terrain was undergoing upheaval new types of animal arose through a feedback process, because they needed the land for further food, and because the waxing and waning ice-ages together with (or as a result of) the early breakup and reformation of the continents often had major and fatal consequences for marine and land faunas. Old ecological niches were destroyed in the process and new ones opened up. Seas receded and new land surfaces arose and joined up with others, inviting colonization of a whole ecological community.

From Sea to Land

But let us look again at the quirky, fortuitous way in which evolution favours particular phenotypes.

It appears that the precocious young of some invertebrates could reproduce without achieving adulthood. They were called the chordates, and started a momentous chain of evolutionary events. They had rudimentary spinal columns and became the forerunners of fish. Original vertebrate material was probably of the softer cartilage type, rather than hard bone, plus a brain-case housing an elaborate nervous system. Even today some creatures, such as modern sharks, still have a solid gelatinous backbone, probably exemplifying an embryonic structure midway between the cartilaginous and bony vertebra of more developed classes. However, the chordates were not quite fish, since they were without jaws or fins and fed by sucking in water and whatever nutrients it contained.

Another important staging post in animal evolution in the seas was a change from the soft-bodied creatures to those with a hard protective exterior, as predatory species became more numerous. They were the first molluscs (such as clams, snails and octopuses) and arthropods (insects and crabs), and were halfway between being single-celled organisms and a complete colony of cells. This change occurred at the beginning of the Paleozoic era starting with the Cambrian period, about 570 to 700 million years ago. Later they evolved into large insects, and fed on the smaller creepy-crawly species then becoming abundant, and on plants and other shallow marine life. From the primitive algae of the Archaeozoic era, which ultimately would continue as mosses and fungi right up to the present day, there was a branching off of the lycopodiates, early ferns, cycladals and filicales.

Once more we are obliged to take notice of biogeographic factors. Rising levels of oxygen would benefit the slowly evolving vertebrates, with mutations involving stiffening through the growth of external coverings, which ordinarily would have been lethal since they would inhibit the diffusion of oxygen. They had a flexible backbone, composed largely of calcium phosphate, with hinges able to support the muscles and other soft parts. They became strong and agile swimmers, still vulnerable to attack but much less so than the invertebrates. These new streamlined creatures became the first fish and soon underwent a further important change when they began to develop a movable jaw, paving the way for the Devonian period — the 'Age of Fishes'.

So far this has taken nearly three billion years of slow evolution, occasionally fuelled by a flurry of faster change. All the while important geophysical events were unrelenting. On the whole the climate of prehistoric times was warmer and more uniform than it is now.[17] The history of the Earth before about 2.5 billion years ago was one of a high rate of heat-loss mainly through vertical tectonic motions. All this implies greater volcanicity because the Earth was attempting to outgas radioactive heat at the spreading ridges, and so massive 'hot spots' appeared.

From Cambrian times onwards temperatures had been alternating between warm, moderate and glacial. As time passed there was a possible lowering of local temperatures in the succeeding Ordovician period. This may have had something to do with the start of the Variscan orogeny. And yet, according to fossilized coral reefs, when the period of orogenis ended the Silurian climate became warmer. Life by that time had appeared on land above sea level, and there were the first plants and terrestrial invertebrates.

Ocean levels were not constant, being dependent on a complex interchange of geophysical events requiring an understanding of the way in which sea and land change place with each other (*see* Chapter 9). Plate tectonic movements and internal melting, gravitational segregation responsible for the distinctive differences of rock type, and contrasting local characteristics all played their part. Spreading ridges can decrease the total volume of the world's oceans, but a prolonged period of volcanism may also have increased the total volume of water.

Some 360 million years ago the beginnings of dense forests showed up in the fossil record, with tall, looming foliage. It was shortly after, during the Carboniferous period when the first reptiles appeared, that really massive tree-growth was achieved. In the meantime the Devonian temperature increased and desert areas expanded. The fish and amphibians had the sea to keep them cool, and the neo-reptiles, 'designed' to cope with drying-up ponds, hibernated in the mud. By the middle of the succeeding Triassic, which was warm and equable, all major fish groups had evolved.

Fig.7. Osteolepsis *was a lobe-finned fish from the Devonian period that was the likely ancestor of the amphibians.*

The Beginning Reptiles

Then another line developed from the main line of succession of bony-jawed ray-fin fish, which had a tougher fin structure useful for sifting through seabed mud. Some, the lobefins, developed armour but in the process became considerably less agile than their softer compatriots. The rays, more flexible and better equipped for swimming, became the most successful; the lobefins much less so. But it was the latter who paved the way for the evolution of vestigial limbs that enabled them to crawl onto the land.

The transition from aquatic to terrestrial eco-systems probably occurred when the climate became increasingly warm. During the Paleozoic there were warmer oceans and the thermal stresses were probably rigorous. It was with the onset of a more variable temperature, and the beginning of climatic zones, in the late Paleozoic, that gave rise to vertebrates attempting to employ primitive temperature-control strategies.[18] By this time much of the swamp forest of northern America had vanished.[19] Shrinking seas caused widespread epidemics and eutrophication, thus hastening the continuing need to get oxygen from the air.

In such situations the survivors in the evolutionary race are invariably the gleaners of oxygen from both water and air. This means they must have been capable of crawling onto the land, like the sole survivors today of such creatures, called the dipnoans, a type of fish with lungs which live in Africa, Australia and South America. They can bury themselves in mud during droughts and live in a cocoon made out of mucus from their skin. By the middle of the Devonian the lobe-finned fish had acquired a suitable set of bones and muscles, and a stumpy fin arrangement that could easily be adapted to a four-footed locomotion. They lived in drying-out river beds, thus getting used to deoxygenating environs.

At the same time it is likely that the insects developed tracheal breathing through perforations in their surface tissue which conducted oxygen directly into their bodies. Tracheal breathing in the open air limits the size to which creatures can grow, and most today have wingspans no wider than 6 cm (*see* Chapter 3).

By the late Carboniferous a vertebrate amphibian with weak stumpy legs had arrived, probably descended from the ichthyostegids. They had a fish tail, jaws with teeth about three inches long, and were the forerunners of reptiles and salamanders. The ichthyostegids in turn came from a group known as the rhipidistias, and radiated (migrated outwards) across the Paleozoic land masses in the moist coal forests with club mosses and giant seed ferns.

It is not clear what happened next, and there is some dispute about how the first amphibians actually breathed. Did the rhipidistias evolve into a creature with an internal bladder, or into one which possessed a diaphragm-type of breathing mechanism? There is an important distinction between the two. Some types of air-breathing fish known as polypterids breathed air not by sucking it in through the mouth and forcing the lung-sac to expand, but by a reptilian-like muscular movement (which is also how mammals breathe). The key to the polypterid method of breathing is the elastic skin of interlocking scales, like body armour, which they used as a kind of diaphragm. As the first land-based creatures would need a tougher, armoured skin to protect them from their new environment, it is likely that these were the first amphibians.[20]

If this is so, it renders rather arcane the controversy about whether the dipnois were the true air-sucking lungfish, or the crossopterygii as is recorded in most textbooks. The view of the American Museum of Natural History, led by the late Don Rosen, is that the dipnois should have evolved into the first tetrapods, or four-legged land-based animals. It does seem however, from the evidence of the Australian National University concerning the Gogo lungfish, that a complicated scenario arose where other techniques of air-breathing occurred accidentally, and that air-sac creatures invaded freshwater habitats while the diaphragm-breathers became the first land inhabitants.[21]

In any event, one kind of fish adapted its fishlike backbones and tougher, firmer fins for crawling with. They became eusthenopteron, a more advanced genus which survived in the fossil record until Triassic times. This further improved their efficiency in the search for more food. They could catch and eat more of their staple diet — worms and insects — surviving in accordance with the laws of natural selection. They were able to make forays onto the land and back for longer periods into the water, to keep their skin damp and to reproduce.

By this time limbs and improved lungs were becoming essential. Massive structural changes were necessary to

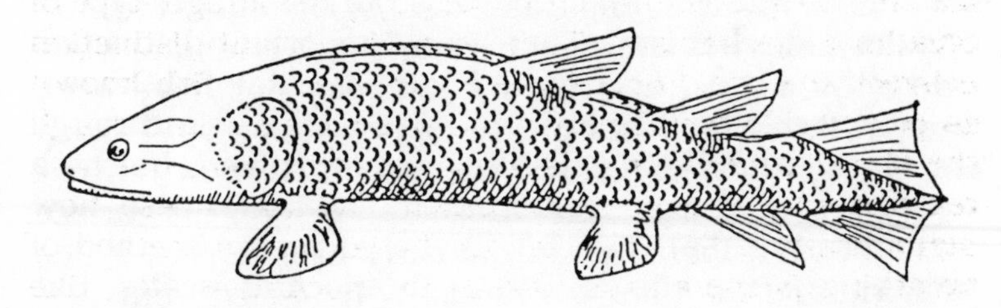

Fig.8. Eusthenopteron, one of the first of the crossopterygii to grow stumpy fins, which later became legs.

facilitate a move onto the land; legs were needed — or at least became handy — to relieve the pressure of the amphibious bodies hitherto capable only of slithering along the ground. And a strong pelvic girdle with links was essential for transmitting the support of the fin/legs to a gradually strengthening spine. Furthermore, those species that accidentally forged some connecting suspension for the front legs independent of the skull would be among the first to walk, otherwise the head would have to turn constantly from side to side as it did so.

The early amphibians spanned the late Devonian right up to the end of the Triassic, some 345 to 195 million years ago, as 'lords of the coal swamps'. The amphibians diffused during the Upper Paleozoic into a number of groups. The last of them, before becoming extinct between the Permian and Triassic periods, were extremely large members of the labyrinthodont group. Most of them had cartilaginous vertebrae, like modern toads. Fossil remains of giant frogs that were coterminous with the labyrinthodonts have been found in the Spanish Tertiary. But they survived long enough for the continents to regroup and to radically alter the atmosphere. They were vertebrates that learned to live out of water paradoxically so that they could also continue to live in water. But more importantly they were the forerunners of a new species that were survivors in a harshly changing Earth. In the meantime, snail-like creatures grew stronger carapaces for protection against drying out. They also developed primitive lungs, as did some lobe-finned fish.

Paleontologists are now certain that vertebrate bone started as a form of armour, and may not have been 'designed' to enable sea creatures to walk on land. Such are the oddities of evolution to which Gordon Rattray Taylor was referring. Bone, in fact, started on the surface of the head, rather than in the early backbones.[22] The other rib and limb bones eventually hardened from their vestigial origins in the cartilage of fishes. Land animals need to protect themselves from drying out, so scales were exchanged for an impervious skin (although some amphibians can control the admission of water through their skin with the aid of hormones). In addition the conquest of the land required a stronger skeleton once the buoyant support of the sea was finally abandoned.

Such embryonic creatures needed stronger muscles, too, and a skin that was resistant to drying out. What undoubtedly stimulated great evolutionary changes in vertebrate history was the emergence of the first amniote egg. Hitherto the ichthyostegids risked the dessication of the unshielded eggs in a dry environment until they had perfected a method of laying them in a slimy, tapioca-like substance. But the amphibians could go one step further, around 395 million years ago, by producing a jelly-like fluid contained within a thin but tough shell, like a miniaturized aquatic environment. It retained water and yet had subtle membranes which permitted air to penetrate and facilitated waste disposal of the embryo.

Thus equipped, these new animals were able to grow bigger and move about freely. In turn the later reptiles could diversify on land when they could lay eggs away from a watery environment. Reptiles differ from amphibians in that, because of the development of the amniote egg, they are able to breed on land as well as live there most of the time. And yet the amniote egg is a curious phenomenon since it arrived, as in other examples, before the environmental need for it arose.[2 3]

In the meantime, ecological stress continued to make its mark felt, and an interrelationship between climatic change and physiological temperature regulation became even more apparent. For example, during the Carboniferous a more bizarre and highly specialized strain of reptile/amphibian emerged — the pelycosaurs. They were large animals that lived at the same time as the forerunners of the dinosaurs, the petrolacosaurs. Their skin was stretched out between long spines, which seems to have acted as a cooling radiator. This is probably the first example of creatures regulating their body heat. Indeed, although such characteristics are commonly thought to be the domain of animals and birds, the evidence shows that thermogenic regulation first occurred in the reptiles. (A pelycosaur sub-order like the *Dimetrodon*, with its great cooling crest, was three and a half metres long.)

The Evolving Reptiles

The pelycosaurs tended to develop rudimentary teeth. Most flesh-eating reptiles have simple spikes that prevent them from chewing their prey; they have to gulp it down whole and then remain in a torpor for days or weeks to digest the meal. With specialized teeth the meat can be sliced into smaller chunks before swallowing it, thus enabling the stomach acids to more quickly penetrate the meal. The evolutionary advantage of this is that the animal need not lie around in a torpid state, vulnerable to attack. And as important as the invention of teeth was, reproduction out of water via the amniote egg was an even more eventful turning point, and in more ways than one. This biologic fact was intricately connected with other climatic and geophysical events that set the mould for future cycles of evolution and extinction.

Indeed, by the late Silurian, colonization of the changing land with its newly emerging continents was well under way. The past was beginning to accelerate, drawing upon what had already been accomplished in its continued drive towards greater complexity, as land surfaces similarly underwent transformation. Both animals and plants were diversifying and were dependent on the increasing amounts of oxygen in the atmosphere.

By now reptile/amphibian morphological distinctions were becoming blurred, and there was a lot of overlap of body shape and biology. The seymouriamorphs were a major phylum on the evolutionary ladder with a skeleton featuring characteristics of both reptiles and amphibians, and probably were descendants of a missing link of transitional animals. The seymouriamorphs appeared in the Upper Carboniferous and developed rapidly and spectacularly into groups of more advanced reptiles, some feeding on plants. Again there was a trend to increasing size, and they ultimately became the predominant land animal, expanding territorially during the last 70 million years of Paleozoic history and preying on the abundance of insects.

Meanwhile important changes were taking place on the land. The climate was turning warm and humid and beginning to dry out the swamplands. Large tropical forests and extensive marshes covered most of the continents. Ferns, club mosses and gymnosperms continued the tree-like growth of the Devonian period, when a moderate climate became warmer,[24] and this dense foliage locked up the carbon as our present coal deposits, thus increasing even further the oxygen content of the atmosphere.

In the succeeding Mesozoic era there was an initial radiation of the reptile class at the start of the first period, the Triassic, with a few more radiations in the following Jurassic. These were all forms which failed towards the end of the Cretaceous, except for the crocodilians which still live on, virtually unchanged, today. In the Triassic there were reptiles that could be classified as either pseudo-lizards or pseudo-crocodiles. The remaining reptiles were probably left-over from the Permian and some were the ancestors of the dinosaurs and other reptiles of later Mesozoic times.[25] Two groups of animals can appear to be distantly related because of their size and body shape, and begin to diverge and then become

Fig.9. An ichthyostegid, a descendant of the crossopterygii from the late Carboniferous, where stumpy fins became stumpy legs.

similar in form much later because both share the same way of life.

By this time species failure was becoming as apparent, from the fossil record, as species survival and progress. Death *en masse* was becoming part of the natural order of things. By the close of the Permian, 75 per cent of all amphibian families and more than 50 per cent of reptile families had vanished from the fossil record. This was certainly more dramatic than the more publicized event that finished off the dinosaurs. One group that disappeared was the fusulinids, complex protozoa that ranged from the microscopic to two or three inches in size. Their shells created thick deposits of limestone. The tiny brachiopods vanished, too. Most of the amphibians had been replaced from the land by about 345 million years ago, and from the water by 195 million years ago, and only the frogs, newts and salamanders now survive.

Episodes of extinction during the past 450 million years seemed to be ecological substitutes and regrouping of better adapted species, but again as a result of retreating seas, by shifting plates and changes in climate.

Fig.10. A seymouriamorph, an important transitional animal in the evolutionary story; halfway between an amphibian and a reptile.

During the Permian period, about 280 million years ago, and until the start of the Mesozoic era, the continents came together in a single mass. Inevitably many marine creatures became extinct when their habitats were literally squeezed out of existence. Species diversity also became limited, thus heightening intra-species competition, with the 'fittest' surviving. These appeared, in the end, to be the dinosaurs. So 'fit' and so well adapted were they, and so successfully did they dominate their terrestrial environment, that they survived for 140 million years.

Chapter 3
DINOSAURS ON EARTH

THE DINOSAURS were the product of the now known laws of evolution. Like other geni and families they continued to change in structure, size and shape during the entire 140 million-year period of their existence. Indeed, so long was their duration on Earth that those species which became extinct at the end were quite unlike those evolving from the reptiles in the early Triassic. The Cretaceous period, in which they thrived but which was also marked by accelerated evolutionary trends that would ultimately make them extinct, is long even on the scale geologists use to measure history — 70 million years. Indeed, the beginning of the period is as far removed from the end of it as we human beings are removed from it in history.[1] In other words, the later dinosaurs were nearer to us in time than their own earliest members.

Dinosaurs probably lived individually for one hundred years.[2] Assuming a generational period of, say, twenty-five years, they would have gone through 5½ million generations — more than fifty times the prehistory of Man. It was inevitable that great changes would take place during the 140 million years, including extinctions of geni and sub-orders, if not entire orders.[3] The fossil record shows that each new generation of dinosaur evinced marked evolutionary changes in body types which set them apart, and which fitted the ecological changes occurring coterminously around them. These changes were mostly in size, gait, metabolism, chewing and digestive mechanisms. In the process, they played a key role in shaping the world as we know it; determining, in fact, the entire course of evolution.[4]

Of course, animals constantly change their form over aeons. If we were to travel back in a time-machine some five million years the physical differences would be quite discernible, even though most species such as monkeys, lions, horses and eagles would still exist. However, 50 million years ago creatures would look much less like they do today; for example horses would have been the size of wolves and have had only three toes. One hundred million years ago was the Age of Dinosaurs: many would look grotesquely large, leathery and fearsome, and be recognizable as dinosaurs to us only because of our clever reconstructions from the fossil record undertaken since the early nineteenth century.

From Crawling to Walking

Our evolutionary story takes us now beyond the proto-reptiles, the lizard-like ancestors of both reptiles and dinosaurs. Some Permian reptiles thrived in a fresh-water environment, preying on more vulnerable victims in shallow waters, such as fish and smaller reptilian types. All the earlier amphibians from the late Carboniferous and the Permian had died out in the Triassic, including the cotylosaurs. Many mammal-like reptiles also died out. Others, however, probably burrowed underground, surviving long enough to become the first true mammals.

We can now discern two major lines: the lizards and the archosaurs, which included the crocodiles as well as the dinosaurs. The archosaurs took over from the mammal-like reptiles about two-thirds through the Triassic. The mammals also existed at this time, starting a little later in the mid-Triassic, but in very insignificant numbers and they were tiny until after the death of the dinosaurs, after which they had more fully evolved from the pelycosaurs and therapsids.[5]

The archosaurs stemmed from the thecodonts, an

order with teeth set in sockets rather than being welded to the jaw as with modern-day lizards. This meant that all the teeth were being continually replaced. As it turned out, socket-teeth were a highly successful innovation, permitting a great improvement in chewing techniques; one of many functional physiological adaptations that would prove to be of enormous survival value.

True crocodilians did not appear until the mid-Triassic, although they probably did not differ much from their late Permian ancestors. In fact, modern crocodiles still have a strong resemblance to their long-lost thecodont ancestors, particularly the 230 million-year-old *Chasmatosaurus*, possessing the same pattern of bony scutes or plates — like armour — beneath the skin. This illustrates yet again the somewhat uneven pattern of evolution, since crocodiles have in a sense regressed by returning to the water.

There were three distinctive characteristics about the archosaurs that paleontologists discovered marked them off from their antecedents. The first was the shape and form of the skull. Living reptiles have a skull like a rigid box, with a simple arrangement of jaw muscles. They are classified according to the number of openings in the skull. The most primitive reptile skull has no openings in the side behind the eye socket (like those in the tortoise and turtles). These were the *anapsids*. In later species there appeared an additional opening behind the eye socket to allow for an attachment of the various jaw muscles; these were the *synapsids*. These applied to the scale-backed pelycosaurs and the mammal-like reptiles.

In the *parapsid* condition there is a single opening high up on the skull. Dinosaurs like the ichthyosaurs and the plesiosaurs, based in swamplands, were parapsids. The largest group of reptiles were *diapsids*. These included all the archosaurs, many types of dinosaur, pterosaurs and crocodilians. Present-day snakes and lizards have them. They have two openings on each side of the skull which made it more like a scaffolding, and accounts for the unusually high number of archosaur

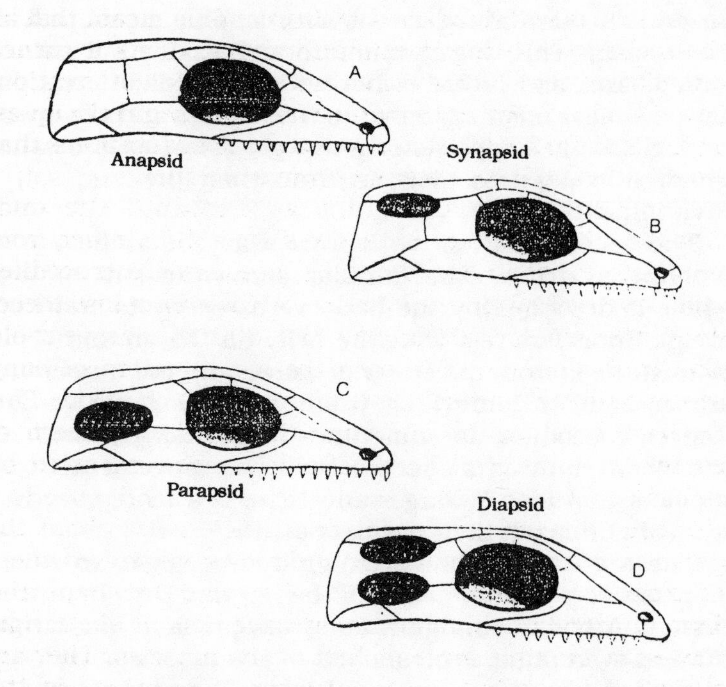

Fig.11. One type of classification of reptiles is based on the number of holes in the skull.

skulls broken into small fragments. One theory about these openings is that they were Nature's way of lightening the head to make it more flexible and manoeuvrable: shattered archosaur skulls might be proof of this.[6]

The largest opening in the archosaur skull was the *orbit*, housing the eyes; behind this were two temporal openings which allowed for the bulging of the jaw muscles and might even serve to make the skull lighter, since with increasing body size every feature that could reduce weight helped.

Euparkeria appeared 225 million years ago at the beginning of the Triassic — a small archosaur. In addition to the other openings it had a curious window-like aperture on the side of the snout with sloping sides, rather like a basin. This was common to archosaurs, but not to crocodilians, and probably housed a large gland. Lizards have a similar opening, between the nostrils and the eyes, for a gland that apparently helps get rid of excess salts which they tend to pick up from their dry and salty environments.

The second distinctive feature was the archosaur's evolving skeleton. The familiar sprawling gait of the reptile is described by the Latin verb *repere* (to crawl or creep). Sprawlers rest with the belly on the ground. The main stride is from the elbow or knee, with the humerous (upper arm or femur) or thigh projecting outwards. Modern crocodiles do sometimes straighten themselves out when running at speed, thus the movement from shoulder and hips is longer and there is a more effective stride, but it is not their normal posture.

The archosaurs, however, gradually began to raise themselves *permanently* off the ground. At first the change in bodily configurations was probably an aid to efficient swimming. *Protosuchus* — about 1.5 metres long with a long head and powerful teeth — was the earliest crocodile. It lived in freshwater lakes in southern Pangaea in the later Permian, and became less of a sprawler, swinging its limbs from the shoulders to the hips, thus lengthening its stride.

Ultimately the hind legs became strengthened at the expense of the front ones. This was a great evolutionary turning point.[7] The tail became long and strong enough for sculling quickly through the water, and the hind legs could give powerful kicks to the river bottom. Here environmental influences again became important, as the early Triassic experienced a drying-out trend that remorselessly began to reduce greatly the number of swamps.

The archosaurs had thus acquired an intermediate position, what has been referred to as 'semi-erect', particularly in those immediately ancestral to the dinosaur. By the middle of the Triassic, thecodonts were diversifying: many by now were able to rear up on hind legs, perhaps to catch insects.[8] Soon the archosaurs became distinctly superior to the reptiles, paving the way for bipedal locomotion and, later, for similarly structured mammals with a more efficient heart and metabolism to come into existence.

Some 230 million years ago a new archosaur, *Ornithosuchus*, was alive. Three metres long and heavy, almost as tall as a horse, it had a menacing array of sharp teeth. The skull was proportionately bigger, narrowing at the front, almost beak-like. It had bony armour along its back, and was able to run in rapid spurts on long legs. Euparkeria, with more pronounced legs balanced by a long tail, was probably a more direct forerunner of dinosaurs than *Ornithosuchus*, but it was smaller and more agile.

The next step, in the late Triassic, was for very many of the archosaurs to become true dinosaurs. Some became known as theropods, or beast-footed, because they walked in an upright (or mammalian) stance, usually on just two feet. This meant that they had now gained a tremendous advantage over the reptiles, even causing the extinction of many of them. They could successfully compete with the true mammals who also evolved in the late Triassic (and who also acquired the upright gait shortly after the archosaurs). Dinosaurs, too, walked on tip-toe (i.e. they are known by paleontologists as being digitigrade)[9] whereas sprawlers pad along on their soles. The anatomical exception was that in the upper region of the dinosaur's femur a hinged joint occurred in the cotyloid cavity, implying that the leg could be moved forwards and backwards. There was also an opening in the three adjoining hip bones, and the upper bone fused to the sacrum had a lateral bony ridge to prevent the femur from slipping out of joint.

The simplified distinction, from then on, between all dinosaurs is that some were bipedal (two-legged), and some were quadrupeds (four-legged). The interesting fact

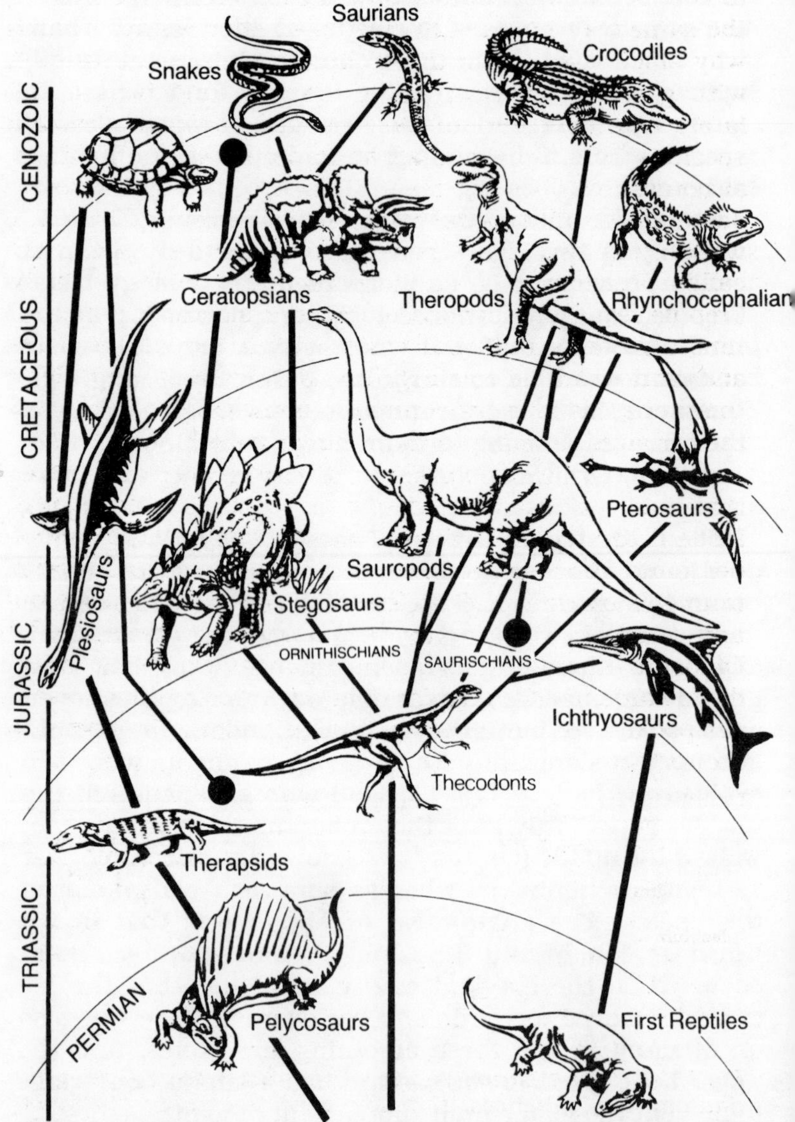

Fig. 12. Shown here are the reptiles from which the two groups of dinosaurs probably originated. (Source: Purnell's Prehistoric Atlas*)*

to note is that dinosaurs soon benefited from bipedality in the same way humans later did — better vision. That is why smell is so important to quadrupeds; erect animals with good eyesight can detect danger from far off. But later, from the end of the Triassic onwards, certain species within the genus began to drop down on all fours, although this does not mean that the quadrupedal forms descended from bipedal forms — or merely got tired of walking on two legs. These new four-legged types were called prosauropods, and preceded the true sauropods ('reptile feet') who lived later in Jurassic and Cretaceous times. Not only that, but where bipeds like coelurosaurs and carnosaurs increasingly had only vestigial front legs, the sauropods (like *Brachiosaurus*) instead began to grow their front legs longer than their hind ones.

Other curious evolutionary changes were occurring that finally distanced the dinosaurs from the thecodonts. Modern lizards and crocodiles have saddle-shaped shoulder joints, concave from the bottom to the top and convex from the inside out. This lets the upper arm swing out and twist; but it is only applicable to sprawlers.[10] Dinosaur shoulder blades had concave sockets facing downwards and backwards. The upper edge of the shoulder pocket overhung the lower edge, more appropriate for an upright stance.

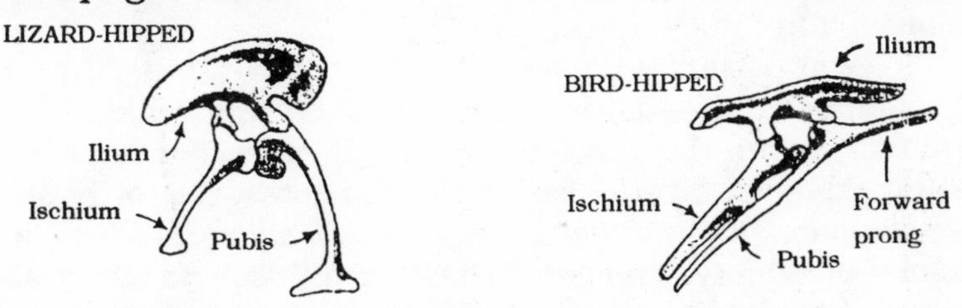

Fig.13. It is the shape of the hip bones, rather than other anatomical differences, that are used to classify dinosaurs. The 'lizard-hipped' types were the saurischians, and the 'bird-hipped' types were the ornithischians. Both types were capable of bipedal walking.

The dinosaur hip socket also had a large hole in the bottom of the basin, unlike the sprawlers.

This meant that not only were there now considerable differences between reptile and dinosaur bones, but there were also emerging differences among the burgeoning dinosaurs. The thigh was straight and had an inturned head, compared (favourably) with that of the reptile which is slightly S-shaped and without a head.[11] The entire body of the dinosaur would have been pivoted around the hip girdle. And it was the configuration of the hip bones, rather than other more obvious anatomical differences, that scientists decided, early on, to use as a classification for dinosaur 'orders'.

Many dinosaurs, those known as 'lizard-hipped', hence became known as *saurischians*. Like other reptiles the saurischians had three bones in the hip: the one above the hip socket is called the ilium, and the one below — pointing forwards and down — is called the pubis. The other bone — pointing backwards and down — is called the ischium. But the anterior of many other dinosaurs was aligned alongside and parallel to the posterior bone (the ischium), instead of pointing forwards as with the saurischians. This bone configuration is today found only in birds, hence the origin of the *ornithischians*. One simpler, broad definition is to distinguish between bowl-shaped and fork-shaped hips.

It is important to stress that both lizard-hipped and bird-hipped dinosaurs were capable of bipedal walking.[12] In fact, soon the majority of fossil types were either slimmed-down bipedal carnivores (coelurosaurs) or larger sauropods, tending both towards herbivory and occasional bipedality. The late Triassic ornithischians were all small bipedal herbivores. Bipedal ornithischians are grouped together as ornithopods, or bird-footed, because of the three-toed feet. It could be argued that the ornithischians were a more varied group. They usually had a thick, armoured skin, as well as possessing advanced features such as predentary bone, and sometimes a complex set of teeth set in a horny mouth. Theropods usually had teeth in the front jaws, and sometimes they extended far back into the mouth. The sauropods instead had teeth in the front with only a few at the sides. The ornithischia often had bony beaks at the front with teeth at the sides and back of the jaws.

The third distinctive feature of the dinosaurs was that many species grew enormously in size. Indeed, this fact is more firmly registered in the layman's mind than any other, and if he or she were asked to give a name to some species, it would probably be either *Tyrannosaurus Rex* (wrongly believed to be the known largest), *Brontosaurus* or *Diplodocus*, the three most often mentioned. *Diplodocus*, in fact, reached 28 metres in length and had an extraordinarily long neck and whip-like tail. *Brontosaurus* was 25 metres long and weighed 30 tons, while *Brachiosaurus* was a long-necked, lizard-hipped creature weighing 80 tons (or more than the weight of sixteen elephants).

T. Rex was a terrible-looking carnivore, some 15 metres long and 5 metres high, as tall as a giraffe and weighing over 8 tons fully grown. It inhabited plateaux and lowlands of the Cretaceous and preyed on duckbilled dinosaurs such as the giant sauropodomorphs. In 1989, paleontologists from the University of Colorado discovered, in Colorado, the remains of an even more ferocious predator than *T. Rex*. Going extinct some 30 million years before *T. Rex*, it was about 16 metres long, weighed about 4 tons and walked crouched forward on three-toed back legs. It had a long, powerful tail like a giant version of *Allosaurus*, the common carnivorous dinosaur whose fossils have been regularly discovered from the Jurassic strata.[13] *Supersaurus* was probably the biggest creature ever to walk on Earth. It was an extraordinary 32 metres long, weighing possibly up to 100 tons. *Supersaurus* is not to be confused with *Megalosaurus* ('giant reptile'), a lizard-hipped carnosaur living in the Jurassic and the very first dinosaur to be named and described, looking like a smaller *Tyrannosaurus*.

Indeed, in many cases some species were even bigger

than we can possibly imagine. Often we have only fragments of bones to build up a mental picture of the final complete skeleton. It is only the natural caution of paleontologists that prevents them from jumping to startling conclusions. For example, Alan Charig and Brenda Horsfield point to the shoulder-girdle and fore-limbs of a likely theropod unearthed on a 1965 expedition to the Gobi Desert in Mongolia. The fore-arm alone was 2.5 metres long, but it was suggested that the creature had disproportionately long fore-limbs, possibly resembling a dinosaurian sloth.[14] This would tie in with the kind of creature that could take advantage of the first abundance of flowering plants and special insects emerging in the late Cretaceous. Even bigger claws half-a-metre long have been found.[15] And in 1972 the matching shoulder blades, vertebrae and pelvis of a sauropod even bigger than *Brachiosaurus* was dragged from the clay in Colorado: the largest vertebra was about 1.5 metres long, suggesting the animal was more than 16 metres tall, weighing over 80 tons and may have been 3 metres in length.

On the other hand, the smallest dinosaur known — of the type that were probably the immediate ancestors of modern birds — was no bigger than a mistlethrush weighing only a few grammes.[15] Many dinosaurs had strange carapaces, like the ankylosaurs and iguanodons and the pachycephalosaurs. The African version of the stegosaur had long spikes all the way down to its tail, with spikes even on its back legs. The short front legs in many such dinosaurs — like the *Diplodocus* and the *Brontosaurus* — reduced the forward weight of the creature. The tail bones were huge, supple and weighty. The base of the tail had large bony flanges which were the anchor-point for immensely strong tail muscles.

The Limitations to Body Size

Curiously, the reason why some species of dinosaur became very large is seldom addressed in either the

Fig.14. Styracosaurus, *a ceratopsian from the late Cretaceous.*

popular or academic literature, and yet as an anatomical fact it can hardly be side-stepped. Dinosaur size is a vitally important subject, bearing as it does on the controversy concerning metabolism and the generation and conduction of heat, and hence on the susceptibility or otherwise to changes in climate. Even Robert Bakker, a paleontologist now at Colorado University and one of the most original and interesting of dinosaur academics, has not, I suggest, given the subject of dinosaur size as much attention as it deserves, and consequently many of his arguments concerning metabolism, which we will examine in the next chapter, remain fatally flawed.

In reflecting on the subject of very large animals, some questions concerning evolution and thermoregulation arise. Why, amongst all the different classes and orders is there such a surprising array of forms, shapes and sizes? What kept insects extremely small over millions of years of evolution, and why have only a few mammals ever approached the size of dinosaurs? Could dinosaurs themselves have grown any bigger?

We ought first to remind ourselves that for every type of animal throughout time there has always been a most convenient size that has fitted it neatly into its own niche. Although evolution decrees that species of plant and animal which survive for long periods are shaped to the advantages determined by natural selection, progress over time to larger size might be a law of nature independent of environmental factors. Richard Dawkins, the zoologist, points out that large size does not imply bigger creatures were intrinsically more successful than species of a small size. There were simply a succession of species of bigger dimensions.

Such species, says Dawkins, were less likely to go extinct than smaller species, or were more likely to split off new, larger species like themselves. The fossil trend towards giganticism was due to a succession of species following the same pattern of progressive development. This would show up in the sedimentary and stratigraphic evidence. So the selection of species could favour a minority of species in which larger individuals increasingly began to interbreed.[17] As we will see, there seems to be a natural trend towards selecting out larger body-size over time, leading to some disturbing conclusions about the likely size of future mammals in millions of years hence.

Meanwhile we should note that one of the important discoveries of zoologists in recent times, one that has tended to blunt warm and cold-blooded distinctions, is the fact that very large creatures, whether animals or reptiles, show much fewer temperature differentials, especially at weights greater than 100 kg. It is a law of nature that metabolic requirements for all largish animals decrease as they grow larger. Large body-size would, in fact, isolate them from their thermal environment because they would exchange heat at lower rates.[18]

Many reptiles have a preferred temperature range within internally regulated diurnal rhythms. In other words, they have become instinctively aware of changes in daily and seasonal variations in temperature. Dinosaurs would have inhabited both daytime and night-time niches determined by regular modes of activity under the control of biological clocks; i.e. the circadian rhythm. One argument says that mammals only began to grow larger when they no longer needed to lead furtive, skulking lives, hiding in crannies to avoid the fearsome dinosaurs. When the dinosaurs at last died off they could, for the first time, roam about increasingly in the daylight hours. Their circadian rhythm became modified accordingly, and they developed a sense of distance that would lead ultimately to a complete transformation of the brain.[19]

There are other natural evolutionary advantages concerning diet and physiological peculiarities, akin to that pertaining to the giraffe and its unique ability to reach the highest-growing leaves. We have seen how the interaction of environment and genes has had a feedback effect, and how the fossils of primitive horses show they were little more than the size of large dogs. Man himself has grown considerably taller during the quarter-million years he has been on Earth.

One visionary speaker at the British Association (science) meeting in 1984 spoke of the likelihood of rats the size of wolves inheriting the Earth in 50 million years time, when Man — in all probability — is no longer on the scene. The wolf-rats would prey on large rabbits, which themselves would grow to the size of deer, to become 'rabbucks'. There could be sabre-toothed prairie dogs, mice the size of foxes, and new species such as the 'gigantelope', and there would be killer baboons.[20]

It would be clear, without needing to elaborate the point, that if bio-evolutionary changes mean some species are excessively small or large beyond or below the optimum, the extinction of that species is likely to occur. This does not mean that, as a survival technique, individual species cannot vary their body size to suit circumstances. The evidence shows that animals can vary their size extremely rapidly in a short space of time when they become, for example, geographically isolated from the main herd. Islands around the world have often hosted odd dwarf forms of mainland creatures: pony-sized elephants once lived on Malta; and it was reported by Adrian Lester of Cambridge University, in a recent *Nature* article, that red deer from the last Ice Age evolved into a dwarf form in less than 6000 years after being marooned on the island of Jersey.[21] Even in normal situations animals can get smaller rather than bigger: the modern tiger, descended from the genus of the sabre-toothed tiger, is now much smaller.

It is at this stage of our discussion that a little physics comes in useful. We know that kinetic energy increases as length raised to the fifth power. A creature twice the size and height of another will have not just twice the muscular energy but 32 times.[22] Many giant dinosaurs, therefore, would have been very successful in battle, having an important bearing on natural selection. Physics also tells us that there is a logical answer to the seeming conundrum of the diversity of species. The smallest insect alive during the Age of Dinosaurs must have been more than a million times smaller than *Brontosaurus*. Insects number countless millions, and perform vital ecological functions. Yet they can only survive in warm or temperate climes, and it would be to their advantage to grow bigger. Even so, it would have been impossible for an insect to grow much larger, say to have a girth more than a quarter of an inch in diameter. If insects were as large as small mammals they would be killed by their own weight. Their legs would need to be proportionately stronger, more like a reptile's or a mammal's, in which case they would need to have the metabolism of either a reptile or a mammal.

One major difficulty for insects is that they do not have a circulatory blood system. This means that they ingest oxygen from the air via fine hollow tubes, known as tracheae. Oxygen travels by means of billions of collisions of gas particles, a process that would be too slow if the molecules had to travel more than a fraction of an inch. For this reason insects are tiny, otherwise they would suffocate. To grow bigger, the invaginations in their surfaces must necessarily become more complex; they would take up more internal space and leave less room for vital internal parts. The only solution to increasing size would be to have oxygen pumped through the body via a circulatory system.

This brings us to the subject of heat convection and heat loss. The smaller the animal the more heat proportionately it loses from its skin because of the increased area:volume ratio. Small mammals therefore do not live in cold countries, and birds migrate south in winter. The shape and size of creatures is determined by the many fundamental forces of Nature. Small animals are dominated by surface forces other than gravity in ways unknown and unknowable to us. For larger animals gravity becomes much more relevant. That is why the surface to weight and volume ratio, so often neglected in discussions of animal biology, is important.

If one were to multiply an animal's breadth, height and length by ten, its surface area would be increased by 100, but its weight by 1000. It would hence require 1000 times as much food and would excrete 1000 times as much waste products, as we are reminded by the distinguished early twentieth-century biologist J.B.S. Haldane in his famous work *On Being the Right Size.*[2 3] It follows also that gravity has a negligible effect on small animals, because their surface to volume ratio is so large: if, for example, a mouse was to be dropped down a 30-metre well-shaft it would be stunned, but would scamper away relatively intact because wind resistance acting on its relatively larger surface would counteract the pull of gravity. A creature any larger, however, would likely be severely injured or killed outright. Similarly an insect walking on the surface of a pond would have gravity counteracted by the surface tension of the water.[24] But once it gets wet the weight of the water may well kill it.

Gravity, in fact, keeps invertebrate creatures relatively tiny. As creatures evolve into larger forms they need strong internal structures to prevent them collapsing into a formless mass. Gravity would soon yield more disbenefits than benefits. If *Brontosaurus* was any taller, a fall could bring about severe head injuries. As it was, the extremely small head of some dinosaurs no doubt reduced the dangers of falling from a great height.

Large dimensions would make it progressively more difficult to stand and move about. For one thing, there is a limit to how strong a leg bone can be in order to withstand great mass. Simply by growing larger, creatures suffer a continual decrease in relative surface area. In other words, volume grows more rapidly than surface. Therefore, although the strength of a bone depends on the area of its cross section, the legs must hold up a body increasing in weight by the *cube* of its length. So bones of giant humans, 6 metres tall (to give a fictional example), would need to be disproportionately thick to such an extent that mobility and agility would be virtually non-existent. Even with bone cross-sections only 100 times bigger than normal they would have reached breaking point, which is induced by stresses of a factor of ten or above.[25] There have long been stories of human giants in legend and mythology. They may have had some basis in reality because bones of a large primate 3.5 metres tall and weighing 800 pounds, found in caves in southern China, date back to over 300,000 years ago at the time Homo Erectus was still alive.[26]

Heavy bones, incidentally, also require a stronger kind of ankle-bone arrangement, to replace the mammalian rows of small bones that slide past each other. Interestingly, many large dinosaurs had a simple, strong ball-and-socket joint between the ankle bones.[27] More interesting still, recent fossil discoveries show that the hip bones of some giant dinosaurs were almost hollow. In the summer of 1988 paleontologists in western Colorado uncovered the 135-million-year-old pelvis and vertebrae of *Supersaurus*. The pelvis alone measured 2 x 1.5 metres in width, and the specimen was not even full grown. The adult animal would have stood 12.5 metres tall and stretched 40 metres from head to tail. But when being prepared in the laboratory, cavities were found in the bones with tiny corrugated bone formations running at right angles. If it were not for the hollow bones the 30-ton creature would have required inordinate amounts of energy, and thus more food, to move around.[28]

Finally, great size also presents problems for the softer parts of any creature. There is, for instance, the problem of what kind of heart would be needed to propel a large body, and the question would arise as to whether, in the case of very tall animals, a higher blood pressure would be required, with commensurately tougher blood vessels, in order to pump blood from the heart to the brain. The internal organs would become more complicated: they cannot simply increase in size because surface area will not increase proportionately, so more intestinal loops and skin tufts may have arisen and, as with the *Dimetrodon*, bony neck frills may become heat-dissipating devices.

How Mobile Were the Dinosaurs?

The traditional image of a dinosaur as a slow, lumbering creature is logically derived from what we understand about great body size and weight. Often this conclusion is backed up by scientists' discoveries of fossilized dinosaur footprints. Most are washed away by rainstorms or floods, but thousands have survived and become preserved in ossified rock, to reappear when the ground has been split open by paleontologists or the weather. R. McNeill Alexander, a zoologist at the University of Leeds, has spent much of his professional life analyzing the movement dynamics of animals. He reminds us that it is easy to make mistakes with footprints: a few years ago, he said when reviewing a recent book on dinosaur tracks for an educational journal, a track appearing to show long, very fast strides consisted in reality of left footprints only, so the real stride length was only half the apparent one.[29]

There is no clear consensus about how fast or slowly dinosaurs moved. As a general rule they appeared to be *slowish*, considering most of them were large animals, but there seems to be a considerable variation between species of greatly differing body sizes. Beverly Halstead, a distinguished paleontologist at Reading University who tragically died in April 1991, said footprint fossils showed that the carnosaurs usually moved around in packs at about two-and-three-quarters mph.[30] The coelurosaurs, he said, normally walked at 5 mph but could run at up to 8 mph. Although this would hardly seem fast enough to catch prey, the smaller reptilian and amphibious animals would also have run at similarly slower speeds. Only the paramammals would have run faster. The larger dinosaurs were hardly quicker, size for size. Big sauropods, for example, show painfully slow speeds of about 4 kph, equivalent to a human's strolling gait.[31] This, said McNeill Alexander, is surprising in view of the fact that some of these animals had hind legs some three metres long.

Of course, the dinosaur specialist has long recognized that *some* of the smaller species were fairly agile, and could have reached considerably higher speeds than the larger carnosaurs. A good example would have been *Saltosuchus*, the so-called 'leaping crocodile'. Footprint evidence, as in the case of a well-known assembly found in Queensland, Australia, shows a herd of small dinosaurs, some no larger than chickens, 'stampeding' (to use Alexander's description) to escape a predatory giant. Other dinosaur track evidence reveals a medium-sized biped weighing about half a ton, seeming to run at 43kph, which is faster than a man but still considerably slower than a racehorse.

Soon, however, a new controversy arose when it seemed to some zoologists and paleontologists that even the large dinosaurs could very well have had a high running speed, at least as high as an equivalent sized modern mammal. Such conclusions were arrived at by studying the fossilized limb proportions and limb-joint morphology of dinosaurs rather than footprint tracks, although some of these tracks, as we shall see, show speeds a lot faster than those Beverly Halstead was talking about.

Many dinosaurs varied in weight because their mass was distributed differently. Their bone density, as we have seen, may have varied greatly. *Diplodocus* was remarkably lithe, weighing only 10 tons in spite of being some 28 metres long. *Brontosaurus* was a more massive version of *Diplodocus*, with not such a long tail (because of the extremely small head of the *Brontosaurus* in comparison with their huge bodies, the name was recently changed — unnecessarily — to *Apatosaurus*, Greek for 'headless reptile'). *Brachiosaurus* was even heavier than *Brontosaurus* and had front legs longer than its rear ones, the opposite of *Brontosaurus*. Most of the sauropods (like *Brontosaurus*) had graviportal (weight-carrying) feet with thick elastic soles, with the foot expanding on tread-impact with the ground, similar to a modern elephant, to keep the animal from getting stuck in soft ground.

For many years paleontologists were obliged to work with the swamp-dweller model of the sauropods. The obvious lightness of the skeleton meant that this kind of existence for an air-breathing semi-aquatic animal would have been particularly inappropriate, because of the buoyancy of their air-filled bodies. It is for this reason whales and hippopotami actually increase their mass to prevent floating. There is also evidence, as we have mentioned before and shall see in the next chapter, of the extensive use of air sacs in sauropods as cooling devices and for reducing mass.[3 2]

Many sauropods, in spite of their ever-growing height spurred on by the need to reach the foliage of trees growing ever higher, had hollowed-out, weight-saving vertebrae and a pelvic girdle described by one writer as being like a suspension bridge, with the massive erect hind legs acting as the down-pillar, and the long torso and rigid tail acting as the cross-span. Indeed, the six-metre long sauropod neck has been likened to a living crane for lifting the head to feed, with up to fifteen vertebrae possessing unique extra articulations in the neck to combine strength with agility. In addition each had a U-shaped bracket in the top of each vertebra which housed a stout cable-like ligament running the length of the neck.

The theropods were carnivores with a lightweight skeleton varying in strength and size. In addition in the Jurassic there were some truly great coelurosaur types with fearsome talons. The eleven-metre long *Allosaurus* — the 'leaping lizard' — weighing up to two tons, as big as a bus, was able to take strides two metres long. *Allosaurus* was assumed from its long hind legs to be a kangaroo-type reptile that actually bounced on its prey.[33] New French findings hint that the tetrapods probably hopped like kangaroos, as their tracks show left and right hind footprints in parallel, rather than alternating, some reaching up to two-metre intervals.[34]

The leaping deinonychosaurs were a sub-order of the coelurosaurs. They hunted in packs, and were individually from two to three metres long. The extra speed and stride that the longer hind legs could attain was hampered by the slow speed of the shorter front ones. So it was more convenient to raise the body up off the ground and sprint for short lengths on the rear legs with the tail, now much thicker and longer, acting as a counterweight.[35]

The evidence, then, from reconstructions of many paleontologists, is that many dinosaurs were far from being sluggish. In fact they were periodically sprightly in the way reptilian orders are. Robert Bakker, however, said they were more mammal-like in movement and stance, and is credited with the 'galloping dinosaur' thesis. Bakker believed that dinosaur speed could, reasonably accurately enough, be worked out from the angle of the limb joints. Rhinos, for example, have more flexure at their joints than do elephants. But horned dinosaurs had more bends in each joint, and were more rhino-like.[36] As time passed, the collar-bone also shrank in size, hinting at the development of a free-swinging shoulder.

Bakker also argued that the brontosaur footprints found in the 1930s in the Cretaceous Texas limestone showed left and right footprints close to the trackway centreline, hinting that they walked upright. He also points out that legs used to regular running have muscles concentrated at the top, with a system of tendons working the leg attached to the bone. The muscle power of the knee can be gauged from a bony ridge, called the cnemial crest, marking the point of attachment for knee tendons. Big gallopers like rhinos have big cnemial crests, and so do buffalos, giraffes and bison. But *Tyrannosaurus* also had massive cnemial crests.*[37]

Critics, however, point out that dinosaur-joint surfaces were not usually smooth and polished like those of mammals. The sockets holding the cartilages were roughened, implying that excessive gristle existed in the knees, which would inhibit fast movement. But Bakker in reply says cartilage is in fact better for absorbing shocks and for building up hydrostatic pressure.[39] Many fast animals have long shanks or calves, compared with thighs, like gazelles and horses. Relatively speaking, says Bakker, dinosaurs had quite long shanks, but not as long as modern-day runners.

McNeill Alexander adopted a different analytical technique, and came up with a much slower dinosaur than Bakker's. He regularly dissects the bones, muscles, tendons and ligaments around the ankle and knees of quadrupeds in order to gain an understanding of the

* On the other hand, one critic has said that the ceratopsian shoulder girdle and forelimbs show that scapular could not have swung in the same fashion as modern mammals.[38]

whole animal's performance. He is in particular interested in what stresses the leg and foot can take. He found that he could calculate, using a fixed relative stride length divided by leg length, the running speed of any animal, past or present. He assumed that all animal bones would be broken by similar stresses, so he took into account the 'peak forces' as approximating to the body weight. These peak forces increase as speed increases, and demand a commensurately stronger skeleton. In 1990 he came up with a middling conclusion: brontosaurs had bones as strong as those of modern elephants in terms of comparative scale. He believed, like Robert Bakker, that as elephants cannot gallop, then neither could the brontosaurs.[40]

On the other hand, recent fossil discoveries hint that dinosaurs occasionally acted extremely violently towards each other, somewhat belying a sedate, lethargic lifestyle. It was in 1989 that Gordon Bell Jr, of the University of Texas in Austin, noticed that around half of the fifty giant aquatic lizards known as mosasaurs showed the remains of healed gashes in their jaws and skulls. Bell and his colleagues now believe these were caused by head-on fights with opponents, who appeared to lunge at each other in much the same way that present-day crocodiles do, but with more apparent gusto, using strong jaws to tear at each other's snouts. The wounds suggest that these dinosaurs were much more territorially inclined than had been thought.

Other dinosaurs, namely the duckbilled hadrosaurs — herbivorous versions of *T. Rex* — were found with fossilized ribs that frequently revealed healed breaks, hinting that they may have been involved in fights or in ritualized trials of strength in which their massive hind feet were used.[41]

One thing becomes clear about large animal size: such creatures must have an abundant supply of food. It had become obvious early on to paleontologists that the dinosaurs, living in a moist and warm environment, in a sparsely populated animal kingdom, had succeeded in

Fig.15. ***Above*** *Elephants have little flexure in their joints, so cannot gallop; neither could the brontosaurs.* ***Below*** *The horned dinosaurs had much more bend in each joint, and must have been rhino-like in gait. (Source: Robert Bakker)*

evolving into the first vegetation-eaters with no competitors. Soon we see further extraordinary co-evolution factors at work. They bestowed upon the dinosaurs tremendous advantages of both large body mass and upright stance suited, perhaps uniquely, to an early unsullied ecosystem of lush vegetation and wide open spaces, a situation on Earth that was never to recur.

The dinosaurs were successful in maintaining a mechanically harmonious relationship between their very long tail and the rest of their bodies. The tail of iguanodon was artificially curved, and this explains why it had a more upright stance than, say, the tiny *Leosothosaurus*. With a straight tail the front part of the body of *Leosothosaurus* has to be lowered, and this levels out the backbone and balances the body around the hips. Iguanodon's curved tail is instead a continuation of its back, with tendons acting as powerful hawsers to counterbalance the front end of the body. This kind of mechanically efficient cantilevered design is more suited to the larger dinosaurs and seems to apply to all the heavyweight bipedal species.

This is not to imply, as has been done with *Tyrannosaurus Rex*, that many large bipedals were little more than lumbering tripods. For the evolving herbivores the advantage of an upright stance was soon complemented by the ability to rear up on hind legs to reach higher-growing plants. *Stegosaurus*, for example, could stand up against the trees to reach upper branches. In this sense (and probably in this stance only) it was tripedal, with the tail acting as an additional balancing limb. *Plateosaurus* was a well-known herbivore, occasionally rearing up and sprinting in short bursts on its hind legs, its momentum aided by its long tail. *T. Rex* and his relatives also had large, curved tails that never touched the ground when the animal was in motion.

This interesting stance has been seen in more recent dinosaur fossil discoveries. John Ostrom of Yale University discovered in 1964 in the Lower Cretaceous of Montana, what he called *Deinonychus* ('terrible claw'). This was a medium-sized theropod, a lightly built, fast-running biped about 3 metres long, with a grasping hand. What Ostrom found, incidentally, was that the second toe of the hind foot did not reach the ground but ended in a large and vicious claw. However, the long tail seemed to be encased in bundles of bony rods which could, like a conjuror's segmented wand held erect by the pressure of the conjuror's fingers at one end, lock it almost rigid. Hence, when the animal was running the tail could have stretched back horizontally and acted as a balancing organ.[42]

The Changing Plant-eaters

The existence of vegetation, and its changing nature, explains much of the evolutionary success of dinosaurs as they adapted to their environment. But it also offers a clue as to their ultimate demise.

The first plants on Earth were a kind of fern, the pteriodophytes, a member of a super-order known as *gymnosperms*. They were of modest size with an underground stem, but without spreading roots. These ferns began to slowly change shape and structure during the Devonian, and some became tall and tree-like. They included conifers much like today's pines, cypresses and sequoias. Other plants were called psilophytes; small, about half-an-inch in diameter, with rhizomes and branched roots. In the Upper Devonian, club mosses and horsetails grew to great heights.

Nowadays *angiosperms*, another super-order, more commonly known as the flowering plants, dominate the world's ecology. Most broad-leafed trees, for example, are angiosperms, as well as grasses, beans, cereals, vegetables and fruit and root crops. Only conifers, cycads, ferns and horsetails today are gymnosperms. Angiosperms grow from ovules (seeds) protected in their own case, where gymnosperms grow only from seeds without the protective case. Botanists now think the angiosperms arose in the Cretaceous, or possibly in the early Mesozoic era, say about 200 million years ago, although possibly much earlier.

Both super-orders throughout time have provided a virtually inexhaustible supply of nutrients for living creatures, which then became prey for the carnivores. The distribution between herbivores and carnivores in nature is often finely balanced. Plant-eaters normally constitute the majority of any population,[43] so the threat of extinction comes when the carnivores start to gain upon the herbivores. Was the destiny, then, of the dinosaurs determined by the herbivores? Did any carnivores die out when their prey had become extinct?

The reason why carnivores exist at all is because they arrived on Earth first. It was probably the archosaurs that started the division between herbivores and carnivores, since all primitive animals were scavengers, or killed prey that killed other prey. The first dinosaurs were also flesh eaters, stalking and killing paramammals and crocodilians. As the dinosaur population began to enlarge, new species of plant-eaters arrived to balance things out. The order of saurischians includes both carnivores and plant eaters. But ornithischians, judging from their fossilized teeth, were invariably herbivores.[44] Today mammals are either herbivores or carnivores, but most plant-eating vertebrates are mammals. By contrast only a few reptiles, like tortoises and iguanas, are herbivores.[45] In nature there is constant pressure to make the herbivores the predominant sub-order. And it is ecologically unlikely that vegetarian dinosaurs arrived because of growing predatory competition among reptiles, paramammals and mammals for live prey.

Things moved more dramatically among descendants of the coelurosaurs. They grew, like the deinonychosaurs, extremely ferocious. They were a sub-order with sharp claws and a prominent second toe looking like a huge sickle.[46] Evidence of fossil tracks shows *Allosaurus* stalked large herbivores, and that it was accustomed to killing and eating animals far larger than itself. This implies that many carnosaurs hunted in herds or packs like wolves, weeding out the old, infirm and unprotected young, since solitary predators like lions tend to kill prey their own size or smaller.

A larger lineage evolved from the small Triassic bipedal herbivores. It emerged as *Camptosaurus* and iguanodon. A study of the skulls of both types showed they probably possessed mammal-like muscular cheeks, and the teeth fossils show they ground grain-like food. There was by now a continual branching off of sub-orders, and more easily acquired food enabled basic structure changes to get under way. Even the danger from predators was not constant until the very end of the Mesozoic. As *Ornithischosus* increased in size its scaly skin replaced the bony armour as the danger from predators lessened. Even so the struggle between the plant eaters and their predators was frequently finely balanced. Often the height and fast bipedalism of the carnosaurs enabled them to tower over the extremely well-protected horned ceratopsians, and simply seize them from above with a powerfully armed mouth.[47] Nevertheless, a combination of structural and morphological modifications, including growing yet bigger and rearing up, mammal-like, to reach higher growing leaves, were great evolutionary advances, integrating important ecological changes. The die was cast for great species longevity.

Indeed, the vastness of the vegetation enabled the plant-eaters to expand, both irradiationally and physically, so that the meat-eaters were seldom deprived. By focusing on the herbivores and their relationship to changing plant life, we can gain a better idea of what happened next to the dinosaurs. Robert Bakker stresses the complete interplay involved here, with sharper teeth to match increasingly 'tougher' plants, and improvements in digestive tracts, playing an interconnected role.[48]

A typical pattern began to emerge. By the mid Triassic there were a few groups of large herbivore reptiles, called rhynchosaurs, who were distant relatives of the lizards.[49] Among the herbivores there were also dramatic changes amongst the bipedal ornithopods, weighing from 30 to 200 kilograms, and ranging from 1.5 to 12 metres in length. These were the bird-footed types. The duckbilled hadrosaurs were an offspring of the larger ornithopods, and started to grow enormous grinding teeth for tough plant materials.[50]

Bakker refers to the Age of Anchisaurs in his book *The Dinosaur Heresies*.[51] Dinosaurs, as well as growing taller, developed a more pronounced downflex to lower their heads closer to the ground for better cropping techniques. Lizards, too, have small, leaf-shaped teeth with edges coarsely serrated; good for shredding but not for chewing. Similarly anchisaurs were long-necked creatures with simple iguana-like teeth, suitable for cropping soft leaves. The cynodonts, with their cheek teeth and incisors typical of carnivores, no doubt preyed on them.

Brontosaurs had spatulate (spoon-shaped) teeth for eating coarser leaves, largely from the ferns and horsetail plants. The *Diplodocus* had the longest neck of all the dinosaurs. It could reach both upwards and downwards to pluck vegetation from all directions on a vast scale, and so lost little time (or energy) moving about. It had peg-like teeth that were usually found in fossils to be heavily worn from perpetually stripping branches of foliage. Bakker calls the reign of *Diplodocus*, *Brontosaurus* and *Brachiosaurus* the 'Age of High Feeders'.

Changing Jaws and Stomachs

There were, however, many extinctions at the Jurassic/Cretaceous boundary, which eliminated most of the high croppers. Only a few of the diplodocuses survived. The fossil evidence seems to support the theory that mammals began to exceed reptiles during the Mesozoic, giving rise to a belief in more efficient eating habits. Some big plant-eaters, like the beaked dinosaurs, did seem to illustrate how creatures often took the wrong evolutionary path; their teeth were too small and less tightly packed in the mouth than the earlier species of beaked dinosaurs.

On the other hand, the soft parts of most dinosaurs took the right evolutionary path. Despite the fact that plants also apparently survived massive quantities of consumption, they counter-attacked well. Many had protective foils like thorns and spines, with much of the plant tissue containing rock-hard phytoliths or tough fibrous cellulose.[52] The 'toughness' of plants is not necessarily a characteristic of either angiosperms or gymnosperms, as it seems that plants struggling to make themselves ever more unpalatable, grew 'tougher' as time passed.

Nevertheless dinosaurs, remarkably, were frequently able to stay ahead in the game. Those that became extinct were replaced by better adapted ones. They probably had very effective digestive systems with detoxifying enzymes to cope with the tougher diets. In animals, digestive systems can either be of the forward or rearward types. Ruminants like deer and cattle have a forward site for fermentation in a complex multi-chambered rumen where the food is soaked and softened by enzymes in the stomach. It is then passed back into the mouth, chewed again, and returned to the rearward locations for further digestion. Other creatures, like some lizards and giant birds such as ostriches, only have rearward fermentation systems in long intestines. They often have tiny heads and no teeth, because the powerful fermentation tracts can obviate the need for more thorough chewing. This would often be accompanied with gastroliths, or 'gizzard-stones'. The gizzard was adjacent to the stomach and lined with hard plates and sandpaper-like surfaces.

Anchisaur species, including modern crocodiles, use gizzard-stones to aid internal 'chewing', to enhance the grinding action of the gizzard. Paleontologists often find a scattered collection of worn pebbles in the vicinity of the fossilized abdominal region. Dinosaurs also had a larger digestive tract that seemed to have co-evolved with the changes in Earth's flora and fauna. The evidence that some dinosaurs had larger, rearward tracts comes from the structure of the abdominal rib-cage. This, incidentally, is another reason why the dinosaurs had to be big. The gizzard-stone treatment meant that food took a long time to digest. As a result much of it had to be stored in a continuously digesting system, so a large torso was needed.

According to Bakker, *Brontosaurus* had short, crowded gastric tracts, with ribs covering the belly-like barrel staves.[53] This may in turn have caused the forelegs of prosauropods (the Triassic forerunners of sauropods) to become more massive, stimulated as they were by the repeated dropping onto all fours when moving slowly. *Brachiosaurus* was a rare specimen, in that it was a much longer four-footed creature, able to consume more vegetation from tall trees with giant forelegs supporting more mass in its digestive equipment.

One problem was that shorter torsos made for better balance when running, but longer torsos could enable longer intestines to become a more efficient non-ruminant digestive system. Bipeds evolved larger digestive tracts which one would normally expect to find in a longer, heavier stomach cavity in front of the hips, which would present anatomical problems in balancing. Dinosaurs overcame this problem by pushing the lower end of the pubic bones behind the hip socket, and so the coils of intestines were thrust backwards from the belly. In the process this helped to balance the forward weight of the rib-cage housing the creature's vital organs.

The evolutionary stage was now set for the third and final phase of dinosaur existence: the 'Age of Low Feeders'. This era was characterized by a marked change in the nature of the Earth's vegetation. The angiosperms started to flourish; 'Coal Age' flora abounded. These new plants could regenerate quickly, so they took over the ecosystem.[54] And they survived the extinction at the end of the Cretaceous, whereas dinosaurs did not.

But this latter phase would have been when the dinosaurs' jaw mechanism was much closer to those of modern mammals, or at least very dissimilar to modern reptiles. The secondary plate, enabling the creatures to chew and breathe at the same time, is also a mammalian characteristic. When dinosaurs' eating habits and teeth and jaw mechanisms are examined more fully, one wonders why only the terminal Cretaceous/Tertiary extinctions are the subject of endless debate rather than any of the preceding extinctions. We must again remind ourselves that there was more than one episode of dinosaur extinction when food and dietary factors have had equal relevance, although each succeeding (and changed) species seemed to be better fitted to an ecologically evolved environment.

Indeed, the jaws of the low-feeding dinosaurs compared favourably with those of modern herbivore mammals. In a horse the bony plate separates food and air passages, the secondary plate reinforces the skull. The horse can move its lower jaw from side to side when eating, and the jaw muscles drag the lower teeth inward in a grinding motion. The muscles operating the system form a complex sling that cradles each side of the lower jaw. This means horses can break up and digest plant food more quickly. Each of the lower feeders had their own cropping techniques, and a general assumption can be made that such techniques had themselves a co-evolutionary impact on plants.

The dinosaurs also developed thick bone where the stresses continually were greater, and holes appeared covered by tough membranes like a baby's skull (known as the 'temporal fenestrae'), where stresses were minimal.[55] In fact *Pachycephalosaurus* had a 10-inch thick skull, and probably used it as a battering ram in ritual contests. But some creatures, notably the American *Allosaurus*, had several bones in its skull loosely joined together so that the entire skull could yield to pressures of various sorts. This enabled it to gulp down large chunks of meat that would have been too large to swallow.[56]

Triceratops is the most famous of the large ceratopsians, and in fact the last to survive. It lived mainly in the savannahs of North America at the end of the Cretaceous. It was more than nine metres long and three metres tall. Its skull — two metres long — was well developed, with three strong pointed horns useful for defence and for stripping some of the remaining shorter fronds, resembling the cycads used in Florida as indoor pot-plants.[5 7] Its horny beak could also dig up roots. The rear rim of the hole in *Triceratops*' skull grew back into a bony frill on which the membrane was attached (although this may also have been a display structure), so that the muscles could enlarge. *Triceratops* maximized its biting strength by evolving high bony cranks on its lower jaws to move the line of muscle pull-up, and thus increase the leverage. Most of the horned dinosaurs evolved even stronger jaw and biting muscles behind eye sockets. The big bony collar could protect vulnerable parts of its body, and could provide a useful broad anchor for muscles which supported the heavy head.

Ornithopods like iguanodon had teeth arranged in interlocking 'magazines' with rasp-like grinding surfaces, lying parallel to one another in the jaw. They could move both upper and lower jaws, unlike mammals or reptiles. Muscles stretching from the brain-case controlled a sideways-moving hinge, which meant that each side of the upper jaw could rotate outwards as the jaws closed.[5 8] Thus opposing teeth would move away from one another as the jaws closed. The jaw movements of some dinosaurs could have a grinding motion like mammals, but most of the mobility came from the upper jaw as it virtually expanded over the lower one.

The last serial dinosaurs were therefore mainly the big beaked ones who fed close to the ground like modern cattle (or more particularly like bison, with their downflexed spinal curvature). Nearly all the Cretaceous herbivores, like the spike-shouldered nodosaurs, were low croppers.

Later came the 'parrot' dinosaurs and domeheads, also with curved downflexed backbones. There were also ankylosaurs, and large and small-horned creatures, and duckbills. In fact the duckbills were traditionally regarded as semi-aquatic creatures, feeding off soft aquatic plants. This belief was due to the alleged webbing of the hands, combined with the horizontal flatness of their tail, supposedly used for sculling. They also seemed to have what could only be snorkeling devices on the tops of their heads. John Ostrom once pointed to the teeth of duckbills that seemed to be specialized for the grinding of coarse plant foods. He observed that the fossils were found in areas populated by both deciduous and coniferous forests, and their beaks seemed to be suited to browsing on trees and shrubs. This theory is supported by the mummified remains of a duckbill which revealed conifer needles, twigs, seeds and fruit of other land plants.[59]

The change in the nature of the Mesozoic vegetation has inevitably spurred questions about a possible link between the success of the angiosperms and the demise of the dinosaurs (*see* Chapter 10). However, the difficulty in deciding cause and effect is illustrated by the fact that the angiosperms appeared after the extinction of the Second Age and coincided with the Age of Low Feeders. There was, of course, a considerable overlap between the gymnosperms and angiosperms. The Cretaceous has also been referred to, for example, as the Age of Cycads. It has been postulated that it was either the dinosaurs that opened up the way for the angiosperms, or instead it was the changing nature of the flora itself that was in some way the prime mover of evolutionary trends; that, in spite of all the advances in jaw structure discussed above, they somehow speeded up trends towards extinction.

The zoologist David Norman points out that the Cretaceous does show a decrease in sauropods, and a rather dramatic increase in ornithopods. He attributes this to the unique chewing methods of the ornithopods, and suggests perhaps it was the flowering plants that promoted the decline of the sauropods and their gizzard-stone type of digestion.[60]

At the very end of the Cretaceous there was a spurt of competition among other herbivores to grow protective

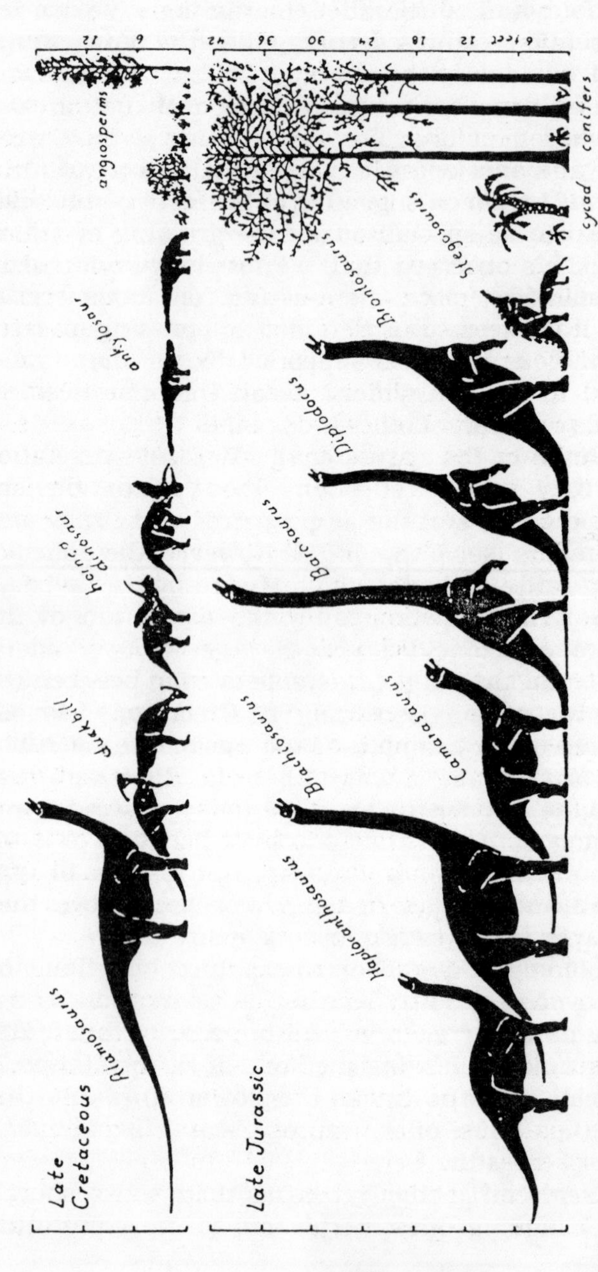

Fig.16. The morphology and size of dinosaurs appeared to change along with the vegetation. The brontosaurs had long necks (lower picture) that enabled them to reach the high-growing conifers prevalent in the Jurassic. Three of the brontosaurs, and a stegosaur, could rear up and feed tripodally. By the time of the Late Cretaceous, only one kind of brontosaur — the Alamosaurus *— existed.* (Source: Robert Bakker, The Dinosaur Heresies)

bony armour and to lengthen increasingly heavy and muscular tails. A lot of them, like the horned and armoured dinosaurs (represented by ceratopsians and ankylosaurs) appeared towards the end of the dinosaur era just when the stegosaurs were passing away. Since a primitive ankylosaur looks somewhat like a primitive stegosaur, it has been suggested that the two sub-orders were really one which split apart later on in the evolutionary story. This contrasts with a later theory which says that the split took place much earlier, say in the earliest Jurassic, if the two sub-orders did not evolve separately from primitive ornithischian stock.[61]

It used to be thought by many scientists that the armoured plates stuck outwards, as in *Stegosaurus*, in order to hit enemies approaching from the side. Others thought they may have been blood-cooling devices, similar to the large ears of African elephants. They were certainly made hideous by the way their fossilized bones, complete with jutting frills, were reassembled in museums. Re-examination of their fossils by Bakker and others show that the thick frills of a *Triceratops*, like the bony scutes of the early *Chasmatosaurus*, was beneath the skin to probably serve as an attachment for the massive cheek muscles and to hold up its heavy head.[62]. Possibly dinosaurs living in the latter phase had developed complex social structures, as today occurs in some reptilian genera. This is likely to have been the case with the final dinosaur group evolving from the small types such as *Psittacosaurus*, or the parrot-type reptiles such as the heavily built quadruped ceratopsians.

In fact more than a million tracks made by iguanodontids, uncovered recently in the Rocky Mountains and dubbed a 'dinosaur motorway', have thrown a new light on dinosaur life-cycles. Martin Lockley, of the University of Colorado, says the tracks show dinosaurs following migration paths similar to the great movements of wildebeest across the Serengeti Plain in Africa. It seems that the herds instinctively followed food sources as the seasons changed in Mesozoic America. The implications were that the dinosaurs were not stupid, slow, cold-blooded animals, said Angela Milner of London's Natural History Museum. There is evidence of a stampede as meat-eating theropods attacked a herd of herbivorous hypsilophodontids browsing in a region of Australia, and some paleontologists have managed to work out the method of attack of the predators and the defensive posture of the herds.[63]

By now, however, in the twilight of their existence on Earth, problems of climate were becoming more important, having finally a crucial bearing on whether they would continue to live on as a species, or succumb. In spite of all their adaptive abilities in regard to diet, armoured defence, body size, chewing and digestive mechanisms, it seems that the 'thermal niche' was the factor that ultimately became seriously disturbed. But this is curious: we are, after all, dealing with 140 million years of almost perfect adaptive survival. Dinosaurs were without doubt cleverly adapted: among the family were insectivores, herbivores and carnivores, generalists and specialists of many varying habitats. In fact they were everything, as Kenneth Hsu of the Geological Institute of the Swiss Federal Institute of Technology at Zurich says: 'from wood chewers to grasshopper catchers, conifer browsers to possum hunters.'[64] Each succeeding threat to each separate family or species was successfully overcome. Dinosaurs could not have been moribund because they were still diversifying into new orders.

We have seen how the findings of Barry Hall and John Cairns have suggested that some genetic codes occur more often when they are useful than when they are not. So the dinosaurs could in theory have continued to pile on genetic advantages. Hitherto it was thought that natural selection is standing still, or merely weeding out the more harmful defects. These new findings might be called Lamarckism. They could eventually contradict the 'central dogma' that changes in dna flow into the protein molecules they code for, but never the other way round. Some critics point out that it is difficult to see how it

Fig.17. The permanently curved backbone of the American buffalo is similar to that of the duckbill species of dinosaur, confirming that they were a browsing animal that had adapted to the changing growth pattern of vegetation. (Source: Robert Bakker)

could benefit larger multi-celled animals, because of the problem of passing on good genes via sperm cells, or the egg. In spite of this one likes to think that the dinosaurs could have engineered their own continuing survival, building upon the immense advantages that large brain and body size confer.

It could possibly have happened before. Indeed, about a million years ago some species of primate — the forerunners of human beings — seemed to select out larger brains in one of the most explosive periods of evolutionary change ever to occur on Earth. The stark conclusion is that environmental forces, as this book will continue to make plain, ultimately defeated the dinosaurs in the race to become intelligent bipeds.

There are important clues to support this conclusion. Largeness, whether as an un-eliminated pituitary defect or as a self-coded dna attempt at prolonging the dinosaurian bloodline, may not have worked to their advantage in the end. The bigger the creature the larger the surface area needed to absorb yet more nourishment. An intelligent brain must possess a minimum mass, although other factors, such as the form and internal neural connections of the brain, must also play a part.[65]

Earth's gravity must, in the end, have been just too powerful. A planet with a small mass, midway between Earth and Mars, would have had a low gravity and a thin atmosphere. Already we could say that tall creatures like diplodocuses had unnecessarily large chests, housing appropriately large lungs needed to inhale an atmosphere that should, in theory, have been thinner than it probably was. Certainly dinosaur bulkiness, dictated by biogeographical and ancestral factors, would have been a continuing dangerous handicap on a planet with Earth-like gravity. We have seen already that the small heads of some dinosaurs with elongated torsos were no doubt a built-in safety factor against crippling head injuries, while naturally restricting at the same time the development of intelligence. One can only assume that the later, powerfully muscled and squatter ankylosaurs were more finely tuned to Earth's gravity. But by this time most of the bipeds — the only creatures able to develop stereo vision, a large braincase, together with forelimbs and an opposable thumb (like *Stenonychosaurus inequalis*) — were already becoming extinct.[66] Also, in an abundant world, there was probably insufficient stimulus to develop intelligence.

Highly intelligent humans, of course, are warm-blooded. Here, perhaps, is the most important clue to the demise of the dinosaurs. Cold-blooded creatures would need very even temperatures if their intelligence was not to switch on and off with variations in the weather. But can we assume they were permanently — that is, intergenerationally — cold-blooded? We cannot, since over 140 million years it is likely that the threat of extinction due to climatic change may have been continually postponed. Dinosaurs were, in fact, a serial species, dying off in relays, sometimes even with 'new, improved' versions emerging from time to time. It is probable that their thermo-regulatory techniques actually went through prolonged periods of metamorphoses.

In the end even that was not enough for continued survival. The dinosaurs probably lost the race against time, having already lost bipedality and vestigial hands at the time when mammalian-type warm-bloodedness would almost certainly have guaranteed their survival into perpetuity. In order to understand why this happened we must now turn, in this final chapter of this section, to the question of how the dinosaurs regulated, or failed to regulate, their body heat.

Chapter 4
WERE THE DINOSAURS WARM-BLOODED?

ROBERT BAKKER brought the subject of dinosaur physiology and phylogeny to the fore of debate in the 1970s. This was partly through his articles, in which he placed as much emphasis on the soft parts as on the hard parts of the anatomy, although the former, for obvious reasons, must be the subject of much more speculation than the latter. Bakker was firmly of the opinion that most, if not all, of the dinosaurs were warm-blooded: i.e. they were endothermic.

Perhaps at this stage it should be pointed out that Bakker has incurred much criticism from his academic peers. He has been accused of indulging in fallacious argument,[1] of selecting data and omitting sources,[2] and he has allegedly ignored the great accumulation of new studies in the 1970s and 1980s that might have refuted some of his arguments.[3] Nevertheless, Bakker has contributed enormously to our understanding of dinosaur biology. Hitherto many specialists opted out of the controversy by saying we have no way of knowing the internal temperature of dinosaurs because of the reasons mentioned above. Bakker has, instead, been ready to use thought and reason to fill in the gaps in our knowledge. He was, in fact, responsible (albeit inadvertently) for goading the American Association for the Advancement of Science into conducting an important symposium on the subject, and his name today continually recurs in most of the dinosaur literature.

Robert Bakker and the AAAS report notwithstanding, animal metabolism is a rather complicated subject, depending greatly for clarity on the precise definition of terms. The metabolism controversy may not have been aided by the somewhat liberal use of generic terms by Bakker and even by his critics, in popular books and articles. So before we examine the debate in more detail let us define our terms.

Briefly, 'warm-blooded' means a thermal state of a creature's core temperature being higher than the low ambient temperature. 'Cold-blooded' means that the core temperature remains close to the ambient temperature 'as it rises and falls'. Here is the scientist's lexicon of formal terms to describe different types of thermoregulation of living creatures, including all living insects, reptiles, amphibians, birds and animals:

Poikilothermic A scientific name for cold-bloodedness.

Ectothermy Where the body temperature depends on the behaviour and heat from the surrounding environment.

Heliothermy The regulation of core temperatures of ectotherms via solar radiation.

Endothermy Where the body temperature depends on high tachymetabolism.

Tachymetabolism A high rate of internal heat production.

Bradymetabolism A low metabolic rate of heat production.

omeothermy or *Homoiothermy* Temperature regulation in tachymetabolic species in which core temperatures remain roughly steady despite ambient temperature changes.

Heterothermy Tachymetabolic temperature regulation exceeding that defined by homeothermy, either daily or seasonally.[4]

If they do little else, these terms show that there is a great risk of over-simplifying the subject when the bald expressions 'cold-blooded' or 'warm-blooded' are used. It is unfortunate, then, that Bakker treats endothermy and ectothermy as if they were mutually exclusive, even suggesting that endotherms have a distinct competitive advantage over the others. As a simple rule of thumb we could say that major differences in thermoregulation are determined by whether heat is derived solely from the environment, or is entirely internally generated, or is somewhere between the two. It should be clear from our list of terms above that there is quite a pronounced divide between ectotherms and endotherms, with the latter having a more varied range of thermoregulatory techniques. Some physiologists insist on making a distinction between temperature regulation and heat production.[5]

In one sense, of course, all creatures, 'hot' and 'cold' blooded, maintain constant temperatures within a certain bearable range, and this means in turn that they must be able to conduct heat away from their bodies. We saw how, in Chapter 2, a biological system of animals functions like any other mechanistic system. An animal must similarly collect 'free' energy and continue to search for new energy sources which must at least repay the costs of the search. One reason why endothermy is not as common in the animal kingdom as one might expect is that the energy costs of endothermy are high. The resting metabolic rate for a mammal is five to ten times higher than for ectotherms with an equally high body temperature.[6]

Living things are always integrated into their thermal environment, and thermoregulation exemplifies stages in progressive adaptation to the rigours of terrestrial life. We could even insist that there is an evolutionary transition not only of skeletal structures, but an evolution from less to more efficient body morphologies. This is because creatures are subject to a range of ambient temperatures within which they can exist at the lowest, and hence most cost-effective, level of metabolic activity, known as their thermoneutral zone. Below and beyond this zone behavioural responses must become the major controlling mechanism, but this becomes metabolically expensive, and must lead to exhaustion if it cannot be made good by an increased intake of food and rest.

Nature requires some compensating feature in any ecological setting for the expenditure and consumption of so much energy. This is why different thermoregulatory devices exist such as skin, fur, feather and scales, all of which play differing roles. Panting, shivering, moulting, all play their part. One clue is that under stress of excessive heat or pursuit by predators, the amount of energy needed or dissipated can be very high. Indeed, food requirements can rise anywhere from five to ten times that needed for a similar ectotherm. We would actually prefer ectothermy in the design of very large animals in tropical climes.[7]

Thermoregulation in Modern-day Creatures

A lot of Earth's creatures, of course, are neither large nor live in tropical climes, and the energy costs of endothermy are not necessarily so high. Let us look first at insects, which are particularly vulnerable to extremes of temperature. When the weather is very cold they cannot fly at all, but can warm themselves by deflecting their wings at right angles to the Sun (as do, for example, locusts). Such would be a *behavioural* response. And at temperatures of 40°C plus, the insect can orient itself parallel to the Sun's rays to reduce the amount of body surface facing the Sun. It can also lever its legs further off the ground, or it can climb vegetation to reduce thermal ground conductivity.[8]

At the same time it can exploit the low thermal conductivity of the air; i.e. the way the temperature of the air itself is neither as hot nor as cool as the land surface. Insects can also lose heat by evaporation from their spiracles, but this may lead to dessication. Rapid flying can soon lead to heat loss by convection, and at night nocturnal insects like moths protect themselves against heat loss with hairs on their thorax which maintain a stable air layer around the body surface.

Let us next look at reptiles. First we must remind ourselves that much of the physiological regulation of warm and cold-blooded animals is determined by the manner in which blood is trapped and channelled to various depths of the body. Reptiles live largely in tropical environments, and the techniques they use for cooling themselves are probably even more important than warming techniques, as we have seen.

Most reptiles utilize the buffering aquatic environment to lower body heat. They have a largely impermeable skin which prevents them from drying out too quickly under a hot sun, although evaporation does take place. Crocodiles can lose water at a rate of up to half of that of amphibians (which have a much more permeable skin, and would greatly suffer from dessication on land were it not for the safety valve of a reduced urine flow, which passes much of its water content back into the body). But the desert lizard, *Sauromaulus*, loses only 5 per cent of that of crocodilians through the skin. Respiration in lizards also plays a part.

Like insects, reptiles are essentially heliothermic, but also have other physiological responses appropriate to ectotherms. The earless lizards of the southwestern United States have a large, blood-filled sinus in their heads which, when they pop their heads out of a cave, becomes heated by the Sun and can be distributed throughout the body. The lizards can then rest on the warm substrate, and seek shelter for the rest of the day.

Reptilian blood systems go through a period of self-adjustment permitting the conservation of heat as required, and allowing the creatures to distribute it when needed. Reptiles become progressively more efficient, just waiting for improvements which only increasing size facilitates, eventually anticipating full homeothermy. Certainly the desert-dwelling species have a surprisingly high and constant body temperature, belying their description as poikilothermics.

Let us now turn our attention to the bird kingdom. We know that birds are warm-blooded; in fact they are fully tachymetabolic, as expressed in the resting rate of oxygen consumption per unit of weight and time. But their feathers also act as insulators. Feathers and (in the case of mammals) hair trap a layer of air which has a low thermal rate of conductivity. Cold-blooded creatures, on the other hand, have fat deposits deeper within the body where they do not interfere with the uptake of heat from the Sun. Birds, also to some extent like mammals, have involuntary nervous controls to regulate heat; for instance, shivering. Nervous controls are not the same as behavioural responses of the kind we have seen in the case of reptiles, which situate themselves at angles to gain the maximum benefit from, or to avoid the worst excesses of, solar radiation.

Most mammals are fully tachymetabolic. They exert themselves to produce internal heat. There is another technique known as metabolic thermogenesis, which is a non-shivering method of producing adrenalin done with the aid of food, in what is known as calorigenic action. This process can speed up metabolism when digestion products are absorbed into the bloodstream. Blood vessels at the surface are constricted in cold conditions so that blood flow is confined to the deeper vessels, which has the effect of conserving blood heat. Similarly when it is hot the superficial vessels expand, so drawing heat away from the interior towards the surface, where it can be cooled by the conduction away from the skin surface.

Biologists talk of 'sensible' and 'insensible' water loss. The former occurs via evaporation and expiration, and the latter through panting and sweating. But sweating depends on the humidity of the air; the higher the humidity the less effective is the evaporation.

Heat loss can also be achieved simply enough by immersion in cold water, which has the effect of displacing the insulating layer of heat. Heat loss occurs through convection, due to the low specific heat content of the air rapidly warmed by contact with the body surface. Warm air rises and is replaced by cooler and denser air, and the cooling effect, of course, is enhanced by winds and draughts. Loss of heat in solid bodies, then, is a function of the difference between the inner core temperature and the surface temperature (i.e. the skin), and can amount to as much as 50 per cent.[9]

Inevitably some creatures, as time passed, endeavoured to find a finely-tuned compromise between the high costs and the regulative benefits of endothermy (the major disbenefit being over-heating). One solution was to periodically suspend thermoregulation; i.e. some creatures became homeothermic or, depending on body mass, became heterothermic. In fact, the ability to hibernate bestowed an ingenious advantage on many creatures. Small homeotherms, like many in the rodent family, do not have the advantage of large size nor of metabolic generation of heat. Most are vulnerable to the cold, since they cannot grow long hair and a thick insulating coat. Homeotherms also shiver. Unlike the head-jutting ectothermic desert lizard, who warms his head first, the hibernating homeotherms would shiver and warm the heart first, then pass warm blood outwards to the body and brain.

The hibernating brown bear is probably heterothermic, with a higher rate of internal heat production than smaller creatures. Hibernation in the marmot, hamster and hedgehog depends on more complex physiological adaptations, and some can drop their basal metabolic rate to as low as one per cent of their most active level.[10]

Assessing Thermoregulation Techniques in Dinosaurs

That dinosaurs, as a sub-order of the archosaurs, are traditionally thought of as reptiles is the legacy of the great nineteenth-century anatomist Richard Owen. He was the man who first coined the term 'Dinosauria', from the Greek word *deinos* ('terrible') and *sauros* ('lizard' or perhaps 'reptile').[11] Owen delivered a famous paper to the British Association for the Advancement of Science in 1841, in which his brilliant insights into the nature of these fearful elephantine creatures and their relationship to the known animal kingdom had a profound impact on the educated classes of his day.

Owen was an early anti-evolutionist. He asserted that dinosaurs, created by God, stood at the apex of the reptilian kingdom, with all descendants of them being vastly smaller and inferior orders. Unlike other giant saurians, he said, dinosaurs were terrestrial rather than aquatic creatures. They differed from reptiles in having five vertebrae fused to the pelvic girdle, [12] with a mammalian-type trunk and a 'thoracic structure' hinting at a four-chambered heart. He was the first scientist to suggest that they may have led a more vigorous life than the reptiles, and even proffered an atmospheric reason as to why they became extinct.

A century and a half later, of course, many new fossil discoveries and much critical analysis have greatly advanced our understanding of dinosaurs, although not to the extent that the controversy and emotion first stirred up by Owen has entirely abated. Argument still centres around their metabolism rather than their physiology, although the latter is frequently invoked to prove some of the assertions made about warm and cold-bloodedness. For example, the paleontologist Loris Russell points out that the skeletal anatomy of dinosaurs appears to be halfway between that of crocodilians and birds. He speculates whether their physiology, and hence their metabolism, might also have been an intermediate one. The fact that saurischians and ornithischians have different anatomies shows that their relationship to each other is as distant as other reptilian sub-orders, like crocodiles and pterosaurs.[13]

But Russell believes that the skeletons of dinosaurs are yet more like birds than reptiles; i.e. he is an Owenite and believes dinosaurs to be more 'advanced' than reptiles.[14] From this Russell concludes that dinosaurs were more like birds in their soft anatomy and physiology.

Nevertheless controversy still centres around the interpretation of dinosaur structure. Bones grow by accumulating crystals of minerals. Dinosaurs did have — and mammals still have — minute channels for blood vessels known as Haversian canals. These tiny grooves, over time, redeposit accumulated bone minerals in concentric layers to release them in a fluid state into the bloodstream when needed in a hurry, say when a violent spurt of activity is needed.[15] Paleontological and zoological evidence shows that ectothermic bones contain few primary vascular channels. On the other hand, as John Ostrom, the Yale geologist points out, Haversian bone in modern creatures is both present and absent in both ectotherms and endotherms. For example, it is present in turtles and crocodiles, and absent in small mammals and some birds.[16]

One argument against dinosaur endothermy is the fact that bones have growth rings in them, somewhat similar to tree rings, especially those found in regions indicative of a strongly seasoned climate. Such rings are hardly ever present in endotherms *nor* in dinosaurs.[17] So the correlation is weak. Beverly Halstead, the late Reading paleontologist, says that studies by Robin Reid have 'utterly demolished the notion that bone could be classified into either ectothermy or endothermic categories.'[18] Yet an argument could be made out for Haversian bone being linked with large body-size rather than thermoregulation techniques.[19]

There are still other hints at warm-bloodedness, derived from recent analysis of fragments of dinosaur eggshells. Scientists excavating the fossilized remains of about twenty nests containing from 1 to 24 eggs each, in southern Alberta near the border between Canada and the United States in 1987, suggest that dinosaurs were more socially complex than was first thought. The baby dinosaurs were an unknown species of hadrosaur of the duckbilled variety. Because the nests are in such orderly rows, scientists at the Museum of Paleontology at Drumheller, Alberta, believe that the same species of dinosaur returned to the site to give birth each year for many hundreds of years.

They also appeared to incubate the eggs, and probably guarded the young; both are characteristics of bird-like activity. Furthermore, comparison with the size of the eggs with the known adult size hinted that the hadrosaurs grew by as much as 280 cm. in their first year after birth. This suggested that they were endothermic, something also pointed out by Robert Bakker, who showed that growth from a ceratopsian egg to a one-ton creature could take just five years. Adult hadrosaurs were up to 13 metres in length and weighed from 4 to 6 tons, about the same as a large elephant.[20]

It has also been pointed out that dinosaur fossils have been found at quite high latitudes, hinting again at endothermy, since no reptile can survive in the cooler climates beyond the equatorial regions. This assertion loses considerable credibility when continental drift is taken into account (*see* Chapters 9 and 10). Nearly all dinosaur localities were situated in what scientists can now work out as being within 40 to 50 degrees of the paleo-equator, where the mean annual temperature was 15°C higher than today.[21]

The other argument concerns the predator/prey ratio. Warm-blooded creatures need more prey; i.e. there is a low predator-to-prey ratio. First, the energy budget of predators is said to decrease with increasing weight, so that lions require ten times their own weight, whereas shrews require 100 times their own weight. Such abioenergetic scaling systems, says Bakker, can cancel out, making body size a constant factor in the predator/prey ratio. He says that dinosaur specimens from late Cretaceous rocks reveal low predator/prey ratios of from 3 to 5 per cent. A herd of zebras, hence, produces about a quarter to a third of its weight in prey carcases per year. A population of mice may yield up to six times its weight because of rapid turnover and high metabolism. The consumption rate of ectotherm predators is equal to its own body weight every sixty days, and this contrasts with the mammalian rate of its own body-weight every 6.6 days for wild dogs, 8.0 for lions, and 10 days for cheetahs.[22]

However, the evidence for this is unreliable, and filtering processes are capable of all kinds of indeterminate errors. Often the burial site is destroyed, or there is a differential representation of habitats. Furthermore, any bias in regard to fossil collectors going for the more interesting and rare predators would actually slant the bias in favour of ectothermy, rather than endothermy. The zoologist Cloudsley-Thompson suggests the enormous size of herbivores 'would be bound to distort the ratio even if the vagaries of fossilization are ignored.'[23] Alan Charig, of London University, also says that no allowance has been made for differing life-spans or even the palatability of the different dead animals, nor for the robustness of fossilized skeletons, nor the latitude at which fossils have been found.[24] On balance, however, the predator/prey theory of dinosaur endothermy does seem to be one of Bakker's stronger points.

Does Bipedalism Prove Warm-bloodedness?

Other issues of controversy focus interestingly on dinosaur stance. Belief in increasing archosaur homeothermy stems from the earlier discussion of the apparent change in posture. It is now commonly held that it was when the upright walkers replaced the sprawlers that the way was paved for homeothermic dinosaurs. The hind legs became stronger and gradually swung in beneath the body to support more of its mass. Would this also require a change in metabolism?

A 'fixed pillar' stance, it is thought, inevitably obliges an animal's system to produce heat even at rest. One writer on dinosaurs suggests that raising the body off the ground requires more energy, and hence a more efficient heart that could supply commensurately more oxygen Bipedal walking and standing, runs the argument, requires the muscles to tense to straighten the legs.[25]

The further the animal's centre of gravity is off the ground, the more work is demanded of the muscles around the joints and girdles, and like all metabolic activity this must release heat.

Hence improved leg performance enabled early archosaurs to be more agile and speedy than other contemporary reptiles, a theory that Richard Owen first articulated. The thecodonts, from this perspective, were one group that improved its locomotion and utilized its energy more effectively, and survived better. Bakker, however, is at pains to point out that dinosaurs were not necessarily built for *sustained* speed any more than mammals are. The cheetah can attain speeds of 97 kmh for about 18 seconds, but soon slows down if his prey eludes him.

Unfortunately bipedality does not prove high activity or endothermy. True, some mammals do sprawl (aquatic kinds such as seals, walruses, dugongs and whales), while no living ectotherm can achieve truly erect gait. Ectothermy does not *cause* slowness; rather, ectotherms simply do not need endothermy. Many are quick and agile in escaping from danger. Even so, reptiles accumulate an oxygen debt if active for long periods, as well as suffering from fatigue and slow rates of repayment.

So all lines of enquiry in regard to body stance are amenable to alternative explanations, perhaps because the issue is looked at from the wrong perspective. It can be maintained, for instance, that erect posture is needed to support a large size rather than represent progress to an endothermic regime.[26] Further, new theories suggest there is some considerable thermoregulatory advantage in being bipedal. Anthropologists, in tracing the evolution of *Homo sapiens* from more primitive quadruped simian species, assumed that bipedalism facilitated the free use of hands from their role in locomotion. *Homo habilis*, using his hands in a variety of building, trapping and tool-making adventures, became rapidly more intelligent. Unfortunately, evidence of either using tools or eating meat does not appear among the fossils until almost two million years later than that indicated when the first upright hominid skeleton was discovered in fossil form.[27]

Neither does the idea of Mary Leakey and her team — that walking upright helped to scavenge dead carcases left behind by carnivore predators — survive close scrutiny, because it appears from hominid fossils that they ate probably more grain than meat.

It was only when attention suddenly focused upon how the earliest hominids kept their large and delicate brains cool in tropical zones that insights occurred. Savanna animals cool off with a kind of organic radiator by evaporating water from the moist linings of the nasal chambers. Panting increases the flow of air through the muzzle and enhances evaporation. This in turn removes heat from the blood flowing beneath the nasal membranes, and this pool of relatively cool blood helps keep the brain from overheating. At high temperatures a powerful pumping movement of the mouth and neck takes place in the large monitor lizard *Varanus*, and is similar to panting in mammals. *Varanus* can also increase its metabolic rate, like mammals but often more effectively.[28]

Similarly, many scientists argue that dinosaurs developed extensive nasal passages with membranes to cool their skin surfaces.[29] John Ostrom points out that most dinosaur nasal passages bypassed the mouth cavity, so chewing could be done without breathing, very typical of creatures with high respiratory rates; i.e. endotherms.[30] The early dimetrodon, as we have seen, did appear to have a cooling device on its back, with the spines helping to increase the creature's surface-to-mass ratio. Other species had radiator-type cranial frills and horns through which blood vessels could dilate and cool, rather like the blood vessels in mammalian horns (such as in stags, antelopes, etc.). These would allow blood temperatures to be moderated with cooler blood from the horn passing into the brain.

Many other dinosaurs, it is suggested, submerged themselves in cooling swamp water, and raked plants with their front teeth, like *Diplodocus.*[31] Archosaurs, the forerunners of dinosaurs, also probably lost heat through their surface skin, or perhaps they had air sacs beneath the skin to effect further cooling, and which also may have made them lighter on their feet.[32]

Fig.18. The large dorsal plates of the Stegosaurus *(above) and the horn-cores of the* Triceratops *(below) could have acted as blood-cooling devices.*

Some archosaur fossils showed that one species, the longisquamata ('long-scales'), possessed heat-insulating scales that possibly also trapped air-pockets and may well have been the forerunners of feathers. Many of the bone-headed ceratopsians could swim with their front feet pawing the lake bed like hippos, both buoyed up and cooled by the water. *Allosaurus* was about 11 metres long and could walk at 5 mph (8 kph), double the speed of the giant herbivores, who could retreat into deep water.

Incidentally, the argument that dinosaurs would have needed to keep their brains cool neither proves nor disproves endothermy. Mammals and birds both have large brains relative to reptiles, and exhibit 'intelligent' behaviour. But they need constant supplies of food and even temperatures. Ectothermic reptiles have a body temperature that tends to vary over a 24-hour period, and hence seem unable to support a large brain. Dinosaurs, traditionally thought to have a small brain, have similarly been analogized with reptiles. However, scientists from the University of Chicago believe that theropods and ornithopods seem to have had brains relatively larger (body mass for body mass) than reptiles, hinting at mammalian-like complex behaviour.

The foregoing discussion of mammalian cooling techniques and their relation to the human species is rather academic. Peter Wheeler, a vertebrates physiologist, reminds us that humans do not have the nasal cooling radiator of mammals.[33] Instead, there are other advantages of bipedalism. Standing upright, he says, naturally reduces the amount of body mass exposed to the Sun's rays in an equatorial landscape. Wheeler did experiments which proved that an animal absorbs 60 per cent less heat at noon in the upright mode than if it was on all fours. A further advantage is that bipedalism raised the body further off the ground, putting more of the skin area in a situation to aid heat loss. Such behavioural modes are not greatly at odds with the thermoregulative techniques of reptilian heliotherms which vary their body stance according to the angle of the Sun's rays.

What Kind of Dinosaur Heart?

Over hundreds of millions of years, as species evolved, they developed better designed hearts with extra chambers which succeeded in separating blood-borne oxygen on its way to the muscles or fins, and back again.

Fish have only a two-chambered heart, consisting of an atrium and a ventricle through which the blood passes on its way to the gills and back to the heart. Reptiles and amphibians use a three-chambered heart, where blood goes into a separate part of the atrium on its way to the muscles, but the blood on its way back empties into the same ventricle. This means that in the case of fish some of the blood must make two or three trips around the lung circuit, and some carbon-dioxide-bearing blood makes several trips round the body before returning to the heart. In the reptilian heart *most* of the blood gets round the body on the first circuit, but with the four-chambered heart — in mammals, birds and (imperfectly) in crocodiles — the blood from the two circuits is not mixed at all. It is as if the body had two hearts — one providing power to the lung circuit, and the other for the body circuit.[3 4] This system gives the animal much more speed and stamina.

Several commentators theorize that dinosaurs may have had hearts like crocodilians, perhaps because they are the closest living relatives to the archosaur. But modern crocodiles possess the imperfect four-chambered heart in which freshly oxygenated blood from the lungs is separated from the used blood returning from the body. This enables oxygen to be stored for several minutes, and during prolonged periods of submersion an accessor valve between the heart ventricles can be opened to bypass the function of the lungs.[3 4] This is supposed to have greater survival value in regard to social functions like nest-egg guarding, and to aid some kinds of brain activity. Such a heart is good, but, say some scientists who believe dinosaurs were more active than hitherto supposed, not good enough.

It is interesting to note that birds and mammals can be

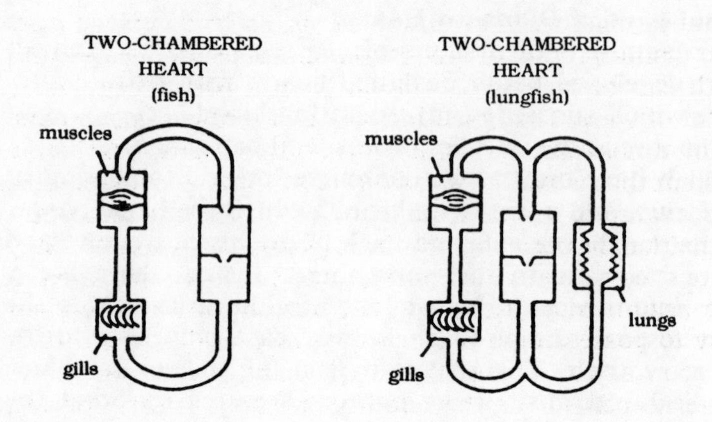

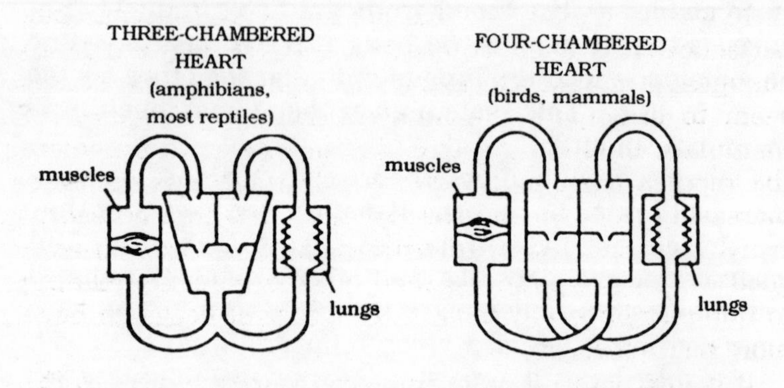

Fig.19. Dinosaurs may have had a four-chambered heart; if so they were probably warm-blooded. (Source: Sprague de Camp, Day of the Dinosaur*)*

grouped together in evolution terms, as both evolved from two distinct reptilian stocks along independent lines; but both can be treated as similar from a thermoregulatory point of view. This itself hints that if birds are descended from dinosaurs, it is likely that the dinosaurs had a system that was different from both birds and mammals. For example, the *Ornithosuchus* is the most important transitional species of the infra-order carnosauria. They were speedy, fast metabolizing, and probably insulated. If the argument is accepted that dinosaurs were on their way to possessing a non-reptilian or bird-like heart, then as soon as the separation of the heart valves was complete it enabled the thecodonts, probably in the Triassic, now with extremely short forelegs, to become lightweight and full bipedal 'bird crocodiles'.

Later changes in the heart mechanism, according to the argument, and increasing size completed the job of further adjustments to their changing environment. But there was soon a limit to this process of adjustment: once perfection had been reached they were vulnerable to changes in habitat. Curiously these metabolic processes seem to anticipate the homeothermic organisation of mammals, thus hinting that dinosaurs carried on where the reptiles left off. In other words there was an ever-increasing trend, with growing mass, towards homeothermy. And once homeothermy had been accepted as a good working hypothesis, then other arguments relating to non-metabolic methods of heat control began to have more relevance.

It is instructive to note that relative newcomers to the dinosaur scene, the mosasaurs (the violent giant aquatic lizards referred to in Chapter 3, who appeared in the last 25 million years of the Cretaceous), showed evidence of mammalian-like illnesses. Mosasaurs, when not head-butting each other, were deep-diving sea hunters, and evidence of avascular necrosis in their bone structure hinted that they suffered in the same way as did human divers — they had frequent spells of the 'bends'. This happens when nitrogen, dissolved in the blood under the high pressure of deep water, returns to its gaseous form, causing bubbles that can block blood vessels and destroy the cells' soft tissue. This never seemed to happen to other diving reptiles such as the ichthyosaurs.[36]

Certainly, according to John Ostrom, dinosaurs themselves showed fossil evidence of mammalian or bird-like behavioural patterns, especially, as we have seen, from footprints suggesting group activity. On the other hand, this view must be balanced by the knowledge that herding and other complex communal activities are also engaged in by a variety of ectotherms.[37] Yet there could be a hindrance to analogizing birds with large dinosaurs because of the question of size (the smaller dinosaurs present an intractable problem).

Comparing dinosaurs with birds provides interesting insights. Birds have a better breathing system than that of mammals or lizards. The left systemic arch has disappeared and the right ventricle empties only into the pulmonary artery, capable of producing large pressure differentials between pulmonary and systemic tracts. It is possible that, over time, the simple loss of the left systemic arch in the crocodilian heart could have resulted in an avian double-pump which would have left the dinosaur 'free to evolve' tachymetabolism.[38] Birds also have a complex system of air sacs through which the air passes before entering the lungs proper. This enables the air to flow in just one direction in an improved exchange of gases between air and blood. In mammals the blood flows in the opposite direction to the flow of air.

The evolutionary and thermoregulatory relationship between birds and dinosaurs was seen to be strengthened in 1989 after an examination of the best preserved fossils of a pterosaur's wings. David Martill of the Open University, and David Unwin of Reading University, discovered the fossilized tissue of the wing of a *Sandactylus* alive 100 million years ago, which had more in common with bat wings than the skin of modern reptiles. Martill and Unwin also found structures that they identified as blood vessels, similar to those found in bats. This enables the bat to cool down as it flies, so if pterosaurs needed to cool down they must have worked their wings as they flew, and probably were endothermic.[39]

The Problem of Body Size

At this stage *body mass* — or size — looms ever larger in importance. It is worth giving it a detailed treatment.

If we say that dinosaurs grew large because they were in pursuit of the advantage of heat conservation, we are not implying that species engineered their own destiny, with the effect predetermining the cause. The thecodonts started off small, so their surface areas were large in comparison with their volume; they could have felt the cold more, and a heat-transmitting skin was a liability. Also an enlargement of the pituitary gland is clearly correlated with large size, almost as if some kind of hormonal imbalance occurred. If this was a genetic accident, then it had tremendous significance in regard to unintentional evolution to homeothermy and great species longevity. This in turn has led commentators to assume that such imbalances and other genetic errors such as endocrine deficiency, led also to their demise[40] (*see* Chapter 10).

Whatever the driving force, heat conservation has many advantages. A study done by Edwin Colbert and his colleagues showed that a tiny 50 gramme (1.76 oz) alligator heated up 1°C every minute and a half from the Sun, while a large alligator some 260 times bigger took seven and a half minutes. Extrapolating this knowledge to a 10-ton dinosaur, they calculated that a one-degree rise in body temperature would take some 86 hours.[41] Halstead says that a 30-ton *Brontosaurus* would take about three days to change its temperature by 1°C.[42] On the other hand, many scientists are certain that gigantism must preclude elevated rates of metabolism.[43]

We must also return here to the issue of dinosaur diet. As the size of a non-ectothermic herbivore increases it has the additional problem of finding enough food (so there is, incidentally, a selective advantage for any plant to become unpalatable).[44] And this is paralleled with the problem, if they were endotherms, of their burning food up too quickly. Indeed, Halstead reminds us that herbivorous dinosaurs would have to eat for more than 24 hours a day! Even if they were ectotherms, a 20-ton creature would need to produce 557 kilocalories per hour. Its surface area would dissipate heat continually at the same rate at which it exceeded ambient temperature, and it would need 8100 kilocalories of solar radiation in addition to its metabolic increment.[45]

Undoubtedly, then, tachymetabolic dinosaurs would have needed plenty of food, and there is some evidence for this.[46] On the other hand, large size for carnivores would not necessarily have meant a remorseless and continued struggle to find food. A single 30-ton sauropod corpse would have lasted *Tyrannosaurus* for three years, leading the Halsteads to surmise that among other anatomical reasons *T. Rex* was a slow-moving scavenger, incapable of moving faster than 2 mph (4 kph). Sprague de Camp also says that the sharpness of the *Tyrannosaurus*' teeth hints that they may have eaten meat softened by decomposition.[47] All cold-blooded creatures move at a leisurely pace, but are capable, as we have seen, of sudden flights of speed when necessary. Remember that, with a very large carnosaur, even a strolling gait would enable it to cover a lot of ground fast and catch fleeing small paramammals without in the least needing to gallop. If the prey were large the carnosaur could not have eaten it all at once. After gorging itself it would lie down and go into a digestive torpor, and could alternately eat and doze over a period of several weeks.

Nevertheless the trend of our argument so far is that evolution seemed to have proceeded towards a faster-moving creature with a different metabolism. There is a revealing Darwinian relationship between size and weight in regard to herbivores and carnivores; one which seemed perfectly balanced to keep the ratio in harmony. But diet complicates an already complex issue. The plant-eaters grew bulkier and heavier as a kind of protection, and in a few instances the flesh-eaters also grew commensurately bigger. Again it depends on how many large carnivores there were in relation to the herbivores. Some plant-eaters were as much as 40 times the weight of *Allosaurus*.

Our ability to gain a better insight into the predatory relationship is made difficult because saurischians could be either herbivores or carnivores. The *Brontosaurus*, for example, was big as well as being a herbivore, and *T. Rex* was big but a carnivore. And yet a similar bipedal dinosaur in size and appearance to *T. Rex* was *Trachodon*, but it was an ornithischian and therefore a herbivore. As a consequence the dinosaurs that preyed on them must also have grown bigger.

One persuasive argument says that many dinosaurs probably needed a high metabolic rate when they were not fully grown, since they had to grow, as we have seen, at a great pace. Halstead implies that dinosaur metabolic rates must actually have changed while they were growing, to convert to the other method completely when maturation was reached. Terrestrial vertebrates with low resting metabolic rates and complete double-pump high-pressure hearts do not exist today. And yet these would be better able to be active and grow quickly on a limited food supply, particularly in a stable equable climate.

The other important point about body size returns us once more to some simple physics. Ectotherms have a very short heart-brain separation.[48] And it is known that the greater the heart-brain distance the greater is the systemic pressure of the blood circulation needed. Crocodiles have low systemic blood pressure, even lower than that of the iguana which has a three-chambered heart. In the case of *Brachiosaurus*, the vertical heart-brain distance was some 6 metres, thus requiring a very high systemic systolic pressure.[49]

This possibly explains the small size of the head of sauropods. We have seen that a small skull for a very tall animal is a protection against severe head injuries. But it has also been suggested that the animal's heart could not have pumped enough blood to nourish a head any larger,[50] or one that was in fact more solid than the light, airy structure of the skull of many known species. If we now turn our attention to the carnosaurs, like the *Dilophosaurus*, and the dome-headed creatures, like the grotesque *Pachycephalosaurus*, we can see that they had very bony skulls. The ceratopsians also had bony heads, seen at its maximum in *Triceratops enrycephalus*, where the brow horns reached extraordinary proportions. This has led to the speculation that these overweighted skulls led to the animal's extinction, as a kind of anti-evolutionary regression.

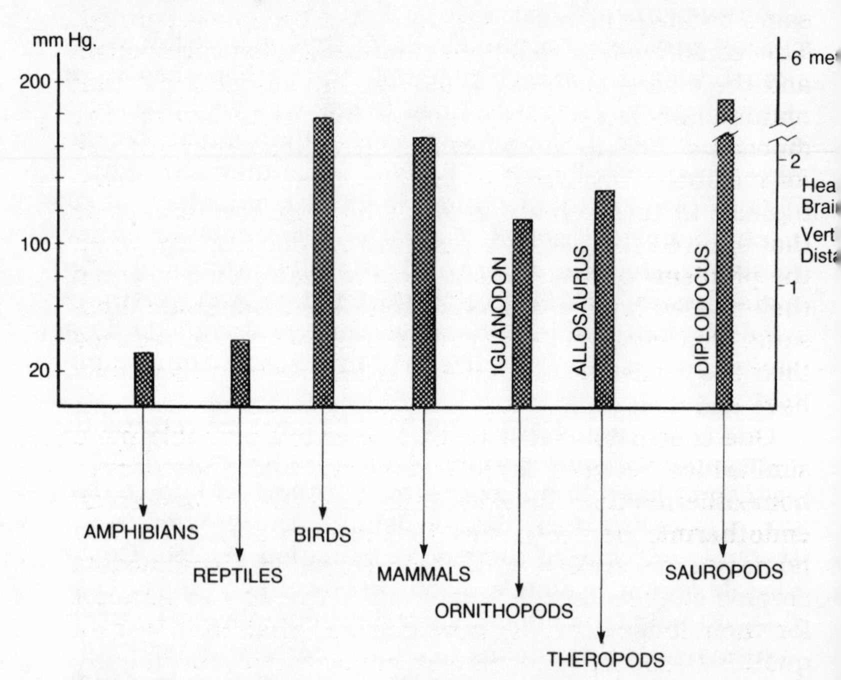

Fig.20. Comparisons of average blood-pressures of modern creatures and dinosaurs. Dinosaur estimates are derived from the distance blood must travel between the heart and the brain, as measured from various mounted skeletons.

Even so, the size-and-weight-of-head argument is a rather specious one. As David Norman has pointed out, the high blood pressure needed in the body and head circuit of the *Brachiosaurus*, with blood being pumped all the way up to its brain along a long neck, would have caused massive bleeding in the thin blood vessels of modern reptilian lungs. Only with the aid of a fully divided heart, with the blood pressure in the lung circuit being lower than in the body-and-head circuit, could *Brachiosaurus* have existed and functioned. Hence it may have been endothermic.

It is possible that during the later Mesozoic small mammals began to co-exist with the much larger dinosaurs because the ecological roles had become separated. This could explain the extinctions: changing metabolism and the emergence of new niches, and competition from mammals better able to survive. In the meantime the dinosaurs' lineage probably reached their optimum size for metabolic efficiency. In fact, it was probably *decreasing* size in those creatures evolving from the therapsids that spurred the need for mammalian endothermy when the problems of overheating became the reverse — when they needed later to maintain temperatures. Some dinosaurs that evolved during the Permian/Triassic reduced their body mass by around 35 per cent or more, and may have had their dimensions reduced by 50 per cent.

One conclusion must be that there are probably more similarities between large ectotherms and endothermic homeotherms than between large endotherms and small endothermic homeotherms like rabbits and mice. But homeotherms are not so well structured to accommodate chronic changes in temperature. This is why to account for their longevity it is now believed that they were a qualified kind of ectothermic homeotherm continually striving to become endothermic homeotherms. One theory suggests that as the non-endothermic therapsids diminished the food base changed, and this in turn obliged the sub-orders like the cynodonts to switch from being passive to active homeotherms. The zoologist J.

Cloudsley-Thompson concludes that dinosaurs 'must almost certainly have been homeothermic.' Bakker, then, in spite of all the criticism levelled at him, was partly right even if he was a little too loose with his terminology.

The problem, as this chapter has made all too clear, is that there are many fine distinctions between homeothermy and full endothermy. Nevertheless, we have seen earlier that homeothermy is a form of temperature regulation in tachymetabolic creatures, and imprecision in terminology has not been the prerogative solely of Bakker. For example, in Halstead's remarkable attack on Bakker in *New Scientist*,[51] he relegated Bakker to an outdated minority of one. 'When the debate [referring to the exhaustive symposium probe by the AAAS[52]] was over, the dinosaurs had lost their high metabolic rate; they were, it turned out, *passively* warm-blooded, inertial homiotherms.' David Norman used his words more precisely when he said they were ectothermic reptiles which were able to keep their bodies at a constant temperature 'by being very large and living in a warm, mild climate.'[53] Without in any way contradicting himself he continues: 'They could have been warm-blooded, highly active creatures (with a fully divided heart), without any of the costs associated with being an endotherm.'

In conclusion we could say that, as finding the correct thermal niche is the hallmark of natural selection, the likely midway position of dinosaurs seems a logical outcome. Extinction hence may have occurred because the thermal niche disappeared. This is where intellectual quandaries arise, complicated by piecemeal evidence leading paleontologists to assume a variety of combinations of thermoregulation. These quandaries can only be overcome, I believe, by remembering how the dinosaur species emerged, and how it died off in relays, and that there was a great diversity of size. Their heads, jaws and teeth evolved with the changing flora. Furthermore, thermoregulation techniques are more flexible than has hitherto been supposed. Natural selection, says environmental physiologist E. Carol Roth, can meet the specific demands of both ectothermy and endothermy in circumstances which are unique to each other.[54]

Roth now believes that animals select what he calls 'optimum thermal surroundings'. They adopt different types of strategies but, as one might suspect, size would require a strategy distinct from other factors. For example, James Spotila says that some lizards are behaviour specialists, flying insects are metabolism specialists, and dinosaurs and elephants are body-size specialists. In co-evolution terms it means that herbivorous dinosaurs may have grown large as a means of attaining thermal stability.[55]

Two intriguing points concern the passage of time. We have seen that perhaps the growing and adolescent dinosaur had a metabolism that actually converted from some form of endothermy to homeothermic or heterothermic metabolism. Nature cares nothing for human definitions, after all, as Sprague de Camp points out.[56] Creatures that bestride the dividing line between amphibians and reptiles and between mammals and protomammals, are constantly turning up in the fossil record. And animal skull fossils sometimes yield tantalizing evidence of transitional metabolic natures. Recently Michel Laurin, a young graduate student in paleontology at the University of Toronto, announced that the lizard-like *Tetraceratops* was an ancient relative of mammals. This conclusion was based on temporal fenestra (or hole-behind-the-eye cavities) that were more like those of the early therapsids. Therapsids were notable mammalian forerunners, and were different from the pelycosaurs in having their jaw muscles attached directly to the skull, rather than to tendons. This gives a more powerful bite, and thus represents an important evolutionary advantage. Laurin concluded that the *Tetraceratops*, as a species, lay between the pelycosaurs and therapsids, and that it must have been on the way to becoming a mammal. The fenestral discovery, incidentally, also filled in a missing link to provide a 300 million year sequence of fossils showing a smooth evolution from reptiles to mammals.[57]

Skin colour, another feature of dinosaurs often neglected, is also of relevance. Spotila, of the State University of New York, reminds us that a white ectothermic dinosaur with a diameter of one metre in a subtropical habitat would have a body temperature ranging from 30.4 to 30.7°C, whereas a black dinosaur's temperature would rise as high as 38.4°C.[58] Spotila cites the findings of researchers who report that the komodo dragon, the largest living lizard, has a body temperature of 36 to 40°C while active. Scaled to size, dinosaurs would have temperatures reaching from 38.5 to 48.6°C, which would imply severe heat stress. Presumably they would be unable to burrow or hide from the Sun like small mammals. So if a dinosaur changed its skin colour from white to black it would cause a 7 or 8°C increase in body temperature (so perhaps colour varied with size?). With air temperatures fluctuating between 22 and 32°C for a reptile one metre in diameter, the body temperature would fluctuate between 28.5 and 29.5°C.

Stephen Czerkas is an amateur American paleontologist who has made a name for himself by reconstructing lifelike models of dinosaurs. Having unearthed the fossilized skin of an entire length of a 10-metre carnotaur (a smaller version of *T. Rex*), he believes he could solve the problem of dinosaur colour and thermoregulation. The carnotaur skin was very ornamental in pattern, with large protruding clusters all over the body, especially along the sides. Czerkas suspected they may not have been, as generally assumed, muddy brown or bivouac green, but daubed with patches of colour around their bodies at strategic spots. They may even have been able to change their colours as do chameleons, to help them attract mates or regulate their temperature. He passed his skin samples to Karl Hirsch, a Denver-based researcher who specializes in fossil eggshells. Hirsch may be able to determine microscopically, as Soviet scientists have in experiments on small mammal-reptiles pre-dating dinosaurs, whether the skin contains tiny gland-like impressions. If the glands secreted sweat rather than scent, that hints at thermoregulation, or warm-bloodedness.[5 9] When the results of these tests are finally known we may at last know for sure whether dinosaurs were warm-blooded or not.

PART II
THE FALL OF THE DINOSAURS

Chapter 5
LIGHT AND LIFE IN THE COSMOS

SO FAR we have seen that the problem of the death of the dinosaurs can now be reduced to one concerning merely the death of the armoured creatures like the ankylosaurs and the ceratopsians. Although carnivores reached their greatest size in the late Cretaceous and were then at their height of development in other ways, overall dinosaur diversity was low.

Dinosaurs never, in fact, 'ruled the Earth': at their peak there were perhaps as few as fifty species alive at any one time,[1] trivial compared with the 5000 species of mammals and reptiles living today. Peter Dodson is a paleontologist at the University of Pennsylvania, and as part of a major survey of dinosaur taxonomy he has attempted to calculate how many dinosaur genera ever lived. He concluded there were no more than 1000, indicating dinosaurs were no more diverse than animals are today. Dodson said he was surprised at this finding, but admits that an expectation of three or four times this number is based on the prolific impression dinosaurs have on the human mind, by their proliferation of Greek names and by the prevailing presence in the world's major museums of their fossilized remains.

In total there are about 2100 dinosaur fossils in the museums. Since 1824 more than 500 genera and some 800 species have been named; but owing to classification errors over the decades, and biases in the fossil record, the fragmentation of the Earth's surface as it broke up and the various extinctions that occurred through the long period that they were alive as a sub-order, Dodson believes that only 285 genera and 336 species can be accepted as valid. Most are represented by a single species, twenty-five have two, nine have three and three have four species. This gives an average of 1.2 species per genus.[2] And according to Beverly Halstead there was no overnight disappearance of the dinosaurs because their numbers had been in decline for the preceding five million years, and only about twelve species were left at the renowned Cretaceous-Tertiary (K-T)* boundary worldwide, although there were rather more species to be found in North America.[3] Throughout the later Cretaceous, in large river valleys, dinosaurs were already yielding to the rapidly diversifying mammals.[4]

The problem with the K-T extinction, more so than with others, is the baffling anomalies frequently uncovered by those investigating it. The mystery is not why so many died off but rather why some species survived at all. The decimation was harsh, uneven and strikingly random. Almost every animal weighing more than about fifty pounds died.[5] Unfortunately there is no simple list of survivors and victims, but we do know that the other species that perished along with the dinosaurs were the curious-looking reptile forms — the flying giants like *Quetxalcoatlus*, together with other warm-blooded flying creatures like the pterosaurs, and the plesiosaurs. The ichthyosaurs were well adapted to life in the sea, but their fossil remains grew increasingly rare as the period advanced.

The ammonites — molluscs with intriguing shells coiled into flat spirals — also became extinct. So did the vast majority of microscopic aquatic creatures. In fact they suffered catastrophically, hinting that any solution

* Geologists identify the Cretaceous invariably with a K, probably to distinguish it from other periods such as the Carboniferous or the Cambrian.

to this mystery will explain all. However, while some larger marine species perished, others (especially deep-sea animals) flourished: by the end of the Cretaceous there were at least twenty genera of these, including crocodiles, lizards and turtles.[6] Species-rich lineages, like clams and snails, fared no better than species-poor lineages, and certainly no better than they did at the earlier, Permian, extinction.[7] The more widely distributed lineages also had no better chance of survival than localized ones.[8] On the land, however, as the Halsteads point out, mammals, birds and non-dinosaurian reptiles as well as plant and insect life remained unaffected. Oddly, the tuatara, the sole surviving member of a rare and pre-dinosaurian group of reptiles, survived.[9]

We must distinguish between *families* of marine animals (of which there were about 790 near the end of the era), and *genera* within families. Families could have any number of genera from one to one-thousand. But whereas, as David Raup points out, the extinction rate for genera was near 50 per cent, for entire marine families it was about 15 per cent. Raup highlights the extinction of many mammal groups, although the lineage as a whole survived. Above all, however, is the puzzle arising from the fact that vegetation was little affected at the end, even if a tentative argument could be made that vegetation in quantity was diminishing throughout the Cretaceous (*see* Chapter 10).

The nearest relatives of the dinosaurs to survive were the birds, perhaps because, it is suggested, they were better able to feed off the decaying matter and seeds in the event of some violent catastrophe that destroyed much of the ecosystem,[10] or because they were home-othermic. When the dinosaurs, already declining as we have seen, finally disappeared from the scene the nocturnal scramblers in the underbrush for over 160 million years came into prominence. Within 15 million years there were bats and whales, and hosts of others.

Energy — the Essence of Life

The previous section, by focusing in large part on the way dinosaurs convert, store and use energy, hints at the kind of approach we will now adopt to the complex problem of their extinction. What I will be suggesting in the rest of this book is that heat and energy flows, not only within dinosaurs but in the solar system at large, had something to do with it.

Let us begin by asking a fundamental question: why did the dinosaurs exist at all? A quite simple (and probably unexpected) answer to this is that they lived in order to fulfil the Second Law of Thermodynamics. The Second Law tells us that energy disperses, and the most common example of this dispersal is the way heat dissipates from a local hotspot into the wider, much cooler, environs. This elegant law of science has a great bearing on the very existence of the dinosaurs. Animal species, once evolution gets under way, seem to become inevitable. But their *continued* inevitability depends on how successfully they function in accordance with the laws of Nature, as these have a bearing on the wider environment.

The Second Law implies that unless they had some way to prevent it, the dinosaurs occasionally got cold. This would not have been their normal state, of course, even if one believes that they were 'cold-blooded'. Our view is that the surviving dinosaurs at the end of the Cretaceous were on their way to becoming warm-blooded. But even if dinosaurs were 'cold-blooded' this does not mean that their bodies either resting or active would have actually been cold, thermometrically speaking, as anyone who has held a small reptile in his or her hand will have noticed. If this had been the case the Second Law would have been seriously violated.

The Second Law is closely related to the Law of Entropy. Things pass from a low entropy to a high state of entropy; i.e. from a low state of disorder to a higher state. They go from a condensed uniform condition to one that is less condensed and less uniform, where types of energy are distributed and exchanged over ever wider areas of the universe. Similarly entropy means that disorder in the universe must inevitably increase, and that energy will continue to dissipate itself into an ever-less available state. Naturally enough the existence of massive and complex creatures seems to negate the idea of a tendency to cosmic chaos and disintegration, but this is only because such creatures existed locally and not cosmically.

The essence of life, strictly, is the endless and complex energy transfers and conversions that sustain the entire ecosystem. As living things are part of the wider universe — in other words, as they are part of a sub-system receiving energy from a larger system — the Law of Entropy, to all intents and purposes, does not apply. Eventually the Earth will die when the Sun's energy is used up, thus finally validating the Law of Entropy. If the dinosaurs had continued to evolve instead of dying out 65 million years ago they would still not have escaped this inevitable fate. Ultimately all the energy in the cosmos will end up evenly distributed as heat, and all further energy-converting processes will come to an end. In the meantime the local alleviation of chaos in the muscles, sinews and blood vessels of the dinosaurs would have been compensated for by *overall* increases in the entropy in the rest of the universe. In any event no energy conversion system can be completely efficient.

The laws of entropy are a scientific abstraction. If the universe hangs together by the grace of radiative force fields, and its component parts are seen to be no more than insubstantial waves of energy, then the universe is almost an abstraction itself, where objects, because they consist of sub-atomic particles, are said to 'collapse into reality' only when they are being observed by scientists. Such beliefs have spawned questionable mystical explanations, which many distinguished theoretical physicists have done little to correct.

For the truth is that energy waves — without any help from the human mind — can manifest themselves very tangibly indeed. Energy can transform itself into heat and even across vast distances can determine the actual temperatures of all planetary objects through gravitational, electromagnetic or kinetic energy forces. A major school of thought, which we will examine in Chapter 7, decrees that the dinosaurs may have died off when a giant missile from outer space struck the Earth. Such an event would be proof that gravity, one of the four force fields of the universe, while preventing our atmosphere from escaping back into space, and literally holding all solid matter together, can be an extremely lethal form of energy, causing collisions to occur in space between hurtling objects sucked into ever-faster orbits.

The impact-extinction theorists believe that a giant dust veil blotted out the Sun's rays for months or perhaps years, preventing most of the Sun's light rays from reaching Earth's surface and in turn bringing about the collapse of photosynthesis. They remind us that terrestrial energy is totally dependent on solar energy. For the moment let us break the argument down into its component parts to see how important is the nature of light to a wider theory of evolution and extinction.

Everything material in the universe is made of atoms, and all energy exists as discrete bundles called quanta. Matter itself is, in a sense, held together by quanta (sometimes known as 'messenger particles'), which it emits and receives from all other matter. One could go further and say that every body (indeed everybody!) emits radiation (known as 'black body' radiation), the wavelengths of which are determined by the surface temperature of the body. Each object will *peak* at a certain range. A warm object peaks in the infrared (IR) band. The wavelengths emitted by matter get shorter as temperature rises. So the Sun, a hot object, naturally emits energy rapidly, peaking in the shorter wavelengths. Indeed, because the Earth is so much cooler it radiates IR energy only over the range of about four microns to over 30 microns.

The dinosaurs would have emitted minuscule amounts of radiation at wavelengths of a centimetre or so in the radio wave sector. They would also have reflected back IR waves which, of course, would have varied according to the depth of their skin colour. Even the action of their hearts in pumping blood around their bodies would have produced electric field variations of fractions of a volt which could have been measured by detectors.[11]

Wavelengths from the cosmos in turn are sensed either by our bodies (such as heat) or by electronic devices and radio telescopes. Without such telescopes astronomers would be severely limited in what they could discover about the universe. Heat energy can be measured either in ergs, calories, joules or as photons, and the energy derived from food can be similarly measured in ergs or calories, as a glance at the label of any commercially packaged food item will confirm. The most important form of quanta — light — is part of the electromagnetic (EM) spectrum, and EM in a sense is electricity in motion. EM constitutes a large part of the physical reality in both stars and planets. It is a long-range force, like gravity, and acts in a similar manner between all material bodies, influencing all particles larger than an atom. Atoms differ in the number of electrically charged particles they possess, and are recognised by the EM interaction taking place between the various subatomic particles, as individual atoms swop and share their outer electrons.

Yet the paradox of EM is the wave/particle duality, where light is seen as first a particle and then a wave. Anything that travels at the speed of light (186,000 miles per second) is more of a wave than a particle, which generally is larger (such as an electron), and slower, but still able to reveal its particle aspects. If, then, light, as quanta, is more of a wave than a particle, then what we perceive as colour is really a wavelength that has been refracted from a beam of white light. Leaves appear green because they have absorbed much more of the blue-red spectrum, and have 'wasted' most of the yellow-green portion. Similar objects appearing red have absorbed most of the yellow-green spectrum.

According to the law of conservation of energy, if a beam of light contains energy then when it strikes an opaque object and is absorbed, most of the energy is likely to be converted into heat. Black absorbs all of the spectrum, and white reflects all of it (including most of the infrared spectrum, which explains why white objects are much cooler when exposed to sunlight than are other, darker, objects). The particles making up the object are likely to begin to vibrate more rapidly, but not quite in the same way that they would if struck by thermal radiation. In other words, these states can be altered by absorbing energetic photons of ultra-violet (UV) rays or visible light.

The Sun — Power Source for the Earth

In the meantime, let us see how the Sun contributes to the total amount of radiation reaching Earth. The Sun is the natural source of most terrestrial energy (some heat energy is generated inside the Earth). The Sun's surface temperature is about 6000°C, so its peak radiation is in the visible yellow spectrum — and it so happens that our eyes perceive this range more effectively than any other.

Under temperatures ranging from 10 million to 20 million degrees C, the Sun's hydrogen atoms are being continually split up into protons, neutrons and electrons. Through an intricate process, some of the protons and neutrons of hydrogen became the nucleus of deuterium (a heavier form of hydrogen), with further fusion completing the conversion to helium. Mass is converted into fearsome heat in the process. The Sun loses mass at a rate of 563 million tonnes a second, but has still only exhausted one-third of its nuclear fuel because its volume is so large that one million Earths could easily fit inside it.

These series of fusion events occur in an uneven staggered manner, with particles being digested and regurgitated endlessly in a cycle that takes several million years to complete. In the process of forcing their way to the surface of the Sun energetic particles and rays are split into their myriad less energetic components, at last producing visible light and heat.

In addition the Sun emits solar flares, which are bright eruptions on the photosphere (the Sun's visible outer surface). It also launches, radially at speeds of up to 500 miles per second, a stream of electrically charged particles. These particles are, in fact, hurled out fast enough to overcome the Sun's massive gravity, taking on average three and a half days to reach Earth (photons, or light particles, take only eight minutes). In 1958 the American physicist Eugene Parker called this outwardly streaming cloud of particles the 'solar wind', which reaches Earth's own magnetic field.

Earlier this century it was discovered from spectrographic studies of sunlight by satellites orbiting beyond our atmosphere, that UV light consists of even shorter wavelengths. UV radiation can be divided into A, B and C Types. UV-A extends from 390 to 320 nanometres (a nanometre (nm) is a millionth part of a millimetre). UV-B extends from 320 to 286 nm, the shortest limit of sunlight reaching the surface of the Earth; but about half the Sun's energy is in the near UV-A waveband, and it is this that reaches the ground. UV-C includes wavelengths shorter than 286 nm, and overlaps the X-ray region at 40 nm.

Laboratory experiments show that oxygen absorbs very shortwave UV-C with peak absorptions at 150 nm, with its absorption decreasing with the lengthening wavelengths. Tri-atomic oxygen O_3, although only about 20 to 50 millionths of the oxygen, absorbs the remaining UV-C radiation, with peak absorption at 260 nm. The 200 to 300 nm band is known as 'far' or hard UV, and is most damaging to life because it has a strong effect on both nucleic acids and proteins.[1 2] Far or hard UV causes most cancers from solar radiation and can penetrate more easily to the ground as the ozone layer thins out.

We saw, in Chapter 1, how the ozone layer helps screen out harmful solar radiation. In addition to ozone, water vapour, CO_2 and oxygen in today's atmosphere also screen out not only UV but IR or visible radiation; molecular oxygen, in fact, filters out lethal shorter wavelengths (10 to 250 nm) of UV. Should the ozone layer ever become dramatically reduced as a result of atmospheric disruption (caused by a bolide striking the Earth, for example), dangerous amounts of harmful UV radiation could reach Earth's surface. Even a small decrease in the concentration of ozone could double the amount of biologically damaging UV coming in.

Sunspots are another complication, also working through electromagnetism. Several American climatologists, like Murray Mitchell of the US National Oceanic and Atmospheric Administration, believe that the 1988 US drought took place about two years after a double sunspot cycle. Another sunspot cycle is predicted for 1991, and other academics point to a period of maximum solar activity in the 1990s. Jim Shirley, based in California, predicts an increase in volcanic activity and more climatic extremes in the near future;[13] and other theorists point to magnetic reversals arising from an asteroidal or cometary bombardment.

The difficulty with so many of these ideas concerns the Earth's magnetic field itself, which tends to vary from one part of the world to another, and from one epoch to another. The molten core of the Earth, some 1370 km in diameter, conducts electrical flows. As the planet rotates we can perceive, with sophisticated measuring equipment, the eddying motions in the core which have the effect of turning it into a dynamo. Possibly the magnetism was created at the very beginning of the creation of the Earth from the collapsing dust cloud, which then became magnetized by the piercing field of the embryonic Sun. Evidence from observation points across the globe suggests that since 1930 Earth's magnetism has increased in many parts of the northern hemisphere. The implication is that 65 million years ago, the solar field, in which Earth's field is embedded, was also producing some abnormal weather effects.

Earth's Protective Envelope

Just why such cosmic and solar radiation arriving at the surface, assuming it is not screened out by our magnetic field or its atmosphere, is harmful, we will discuss in the next chapter. In the meantime we need to know a little more about Earth's protective atmospheric envelope. First, its several layers, in which countless chemical reactions take place. Some of these reactions are initiated by Man's polluting activities, but others occur in purely 'abstract' conditions.

An astronaut travelling upwards from the surface would pass through what we know as the 'air', the densest part of the troposphere extending up to about seven miles. This regions contains more than 75 per cent of the entire mass. Above seven to ten miles he would ascend into the stratosphere. At the lower boundary the temperature is very low, with screeching hurricane-force winds. As he continues to travel upwards the temperature gradually rises, the opposite of what happens in the troposphere below, where the temperature drops constantly by about 1°C for every hundred metres gained in height. But as the ozone layer at the top of the stratosphere is warming it up, the hot air is reluctant to rise further and instead assumes a mode of stratified stability — hence its name 'stratosphere'.

The air is by now far too thin to breathe. In fact it is not much denser than that on Mars. With the higher temperatures on Earth combined with very low pressures, it would also be inimical to life. Above the stratosphere we come to the ionosphere. At this height the action of the Sun and the unfiltered nature of the gases would be dangerous to living creatures, with X-rays and UV wavelengths of less than 100 nm getting absorbed.

Chemical reactions abound, with most molecules being dissociated into single atoms, and some atoms being smashed into their alpha and beta particles, or into positive and negative elements. This portion of the atmosphere becomes 'ionised', hence the name 'ionosphere', and one of its well-known characteristics familiar to radio buffs is its ability to bounce radio signals back from Earth.

Finally we come to the outermost and thinnest layer of air, if we can still call it that. This is the exosphere, which contains only a few hundred atoms per cubic centimetre. Scientists travelling in instrumented balloons to the upper reaches of the atmosphere have noted that all the relative proportion of the gases remain the same, but become less concentrated. Only in the layer of atmosphere within 10-50 metres do minor variations appear, due mainly to the influence of organic life. Sharp variations occur with water vapour content (especially over large continental land masses furthest away from oceanic-based moisture), but this moisture content rapidly declines beyond a height of 60 miles (100 km), where the atmosphere changes completely under UV radiation.

The Sun is a low entropy region, and transmits a lot of its power to the Earth in the form of high energy protons, because there are more degrees of freedom in the greater number of low energy protons than in the smaller number of high energy photons in the Sun.[14] Plants convert these more powerful photons into low energy ones which the first animals ate, and which eventually the dinosaurs ate, and so on down the line.

Clearly the amount of solar power received in the past was directly proportional, other things being equal, to the amount of energy it gave off. A one-percent change in solar brightness could approximate to a one-percent increase in Earth temperatures because most of the radiation is heat and light. In fact most, about 60 per cent, of the Sun's radiation is in the IR region, and about 40 per cent is in the visible region. Only about 3 per cent peaks in the UV region. Taking the atmosphere as an envelope surrounding the Earth, we can say that it is reasonably transparent only to visible light and to microwaves. Other portions of the spectrum are almost entirely absorbed long before they have passed through the air.

In fact solar radiation reaching Earth varies with the

time the path of the Sun's rays takes to pass through the atmosphere either vertically or at an angle, and the changing distance from the Sun which varies seasonally because of the Earth's elliptical orbit. In addition, of course, the Earth is tilted 23½ degrees on its axis, and this has other effects. The northern hemisphere receives more radiant energy during the period 21 March to 21 September, because the angle of incidence of sunlight is more nearly vertical during this period.

Still, Earth's radiation is remarkably balanced by some fine physical constants. Our planet has highly efficient techniques for absorbing energy from space through complicated warming and cooling air circulation patterns. Strange feedback loops occur which can alter the climate, depending upon a complex interplay of trace gases, atmospheric moisture and cloud density, heat and inversion layers. Earth still gains more heat than it loses because of what is known as the greenhouse effect — the way heat is literally trapped by atmospheric gases.

Today the unpolluted atmosphere consists of 78.1 per cent nitrogen (N_2), 20.9 per cent oxygen (O_2), 0.9 per cent argon (A), and very small amounts of trace gases such as neon (Ne), helium (He)m, krypton (Kr), free hydrogen (H_2) and methane (CH_4). It is important to bear in mind that, with an oxygen content of nearly 21 per cent, our atmosphere is at a safe level for the sustenance of life. Even a smaller increase could lead to widespread forest fires. Indeed, for every one-percent rise in O_2 the chances of fires being started by lightning flashes increases by 70 per cent. Over 25 per cent of the world's forests would be devastated with great frequency.

Reaping the Light Harvest

We saw in Chapter 2 that the laws of physics have taught us that gathering solar radiation is a way of harvesting order, although it is clear that the early unicellular organisms must have been in some way immune to the harmful spectra of radiation before the protecting atmosphere came into being.

Life-forms ultimately depend on an energy source to construct their vital organic molecules from simpler ones. Even harmful energy was beneficial when it periodically eliminated competitors or predators. Biological systems similarly harvest heat, either from the Sun or from the Earth itself (often in the form of geothermal energy). Heat continually flows across the globe from different parts of the environment and between the ecosystem and organisms, first one way and then another.

The first primitive organisms lived on chemical compounds in the primeval soup that had absorbed the stored energy of UV rays. Then energy was liberated through the process of fermentation. Many primitive microorganisms today are still utterly dependent on this kind of solar nutrient. But even higher organisms resort to fermentation as an auxiliary process when the oxygen supply is deficient. Today's blue-green algae can carry out photosynthetic process by fixing CO_2 into organic compounds as well as nitrogen, without having nitrogen in the soil. Those creatures and organisms that do not use photochemical energy to survive instead derive their chemical energy from those that do, as can be expressed in the following formula:

inorganic compounds + sunlight + oxygen → organic compounds

Ultimately, at the beginning of the Silurian period, the demand for nutrients became more than could be supplied by the Sun. In this sense they seemed to begin to alter their own environment through their metabolism. New and complex processes of energy production took place, until aerobic oxidizing ultimately occurred as the predominating energy-using form. Oxygen is a very potent source of energy, more so than UV radiation alone can provide. It was inevitable, then, once the necessary 'errors' in mutation had occurred, that oxygen would rapidly pave the way for the development of chemical energy that enables animals to move rapidly. They could more easily retreat from danger, and could think (the dinosaur brain would have needed very high levels of oxygen in order to function properly and to evolve).

We can summarize by saying that as a result of evolution, organisms have beneficially adapted to the Sun's rays in unique ways. Plants glean the energy from sunlight to control flowering. This is known as the photobiological response. Yet chemical reactions take place even more briskly with heat flows than with photosynthesis, especially where water is the solvent. Life can exist only within about the 300°C temperature range — say, from -200°C to 100°C, although the vast majority of species are restricted to a much narrower range. Most life processes cease, for example, when the temperature drops to 0°C, and halts again for most life-forms at around 40°C. For plants in northern latitudes, the optimum temperature range is between 20 and 30°C, and the minimum is around freezing (0°C). Plants living in extreme conditions, such as lichens, have minimum temperatures well below zero, while desert plants can survive from 35 to 50°C. Some micro-organisms can exist and thrive in permafrost, whereas some, chiefly bacteria and algae, thrive in hot springs where the temperature is close to boiling point. Here the upper limit would be 88°C, compared with 50°C for most fish and insects. The range of vegetation is naturally much less in ocean waters, and accordingly aquatic creatures have a narrower range of tolerance to temperature variations.

Nevertheless, Nature's conditions are both inhibiting and regulatory. Life adapts, and falls into line with the periodicities found in Nature as the earth turns on its axis while orbiting the Sun, so that diurnal tidal and seasonal rhythms occur — what is known as the photoperiod. These rhythms trigger physiological activities, such as the growth and flowering of plants, or the hibernation of animals, as well as the migration of birds, and the moulting of furry animals or others who instead 'put on weight'.

It is the seasonal factors that constantly change, and

locality can in many cases change by the cyclical migration of species. Latitude plays a big part, too. The further north or south you go the greater is the variation between the maximum photoperiod in the summer and winter, as a Londoner visiting Oslo will soon become aware.

This kind of environmental-biological periodicity is sometimes known as the biological clock, a form of hidden timing mechanism. Using a kind of sensory perceptor either in the eye of an animal or a special pigment in the leaves of a plant, a living thing can measure, or is sensitive to, time as reflected in the length of day (known as the circadian rhythm). Florists, in fact, can often force flowers to bloom out of season by artificially altering the photoperiod. This photoperiodicity is closely connected to another feature of Nature: the nitrogen cycle. Research has shown that the root nodules of some vegetables fix nitrogen according to the photosensitive pigment of their leaves, thus increasing the amount of nitrogen-fixing bacteria in the soil. Earth temperatures are important, too, with temperature rhythms controlling or greatly modifying other seasonal and daily rhythms. It is often a very significant limiting factor, thus showing how important is the ecological connection between solar radiation and its penetration, and life on Earth.

Chapter 6
DEATH BY COSMIC RAYS?

THE EVOLUTIONARY story of Earth is one of continual triumph over adversity, with victory being snatched time and again from the metaphorical jaws of defeat. The first of such victories took place remarkably early, when there was a literal shortage of prey for microbes in the primeval soup. The solution was to become self-parasitic, a distinctively crude illustration of the lack of any master design or blueprint in the story of unfolding life on this planet.

The procaryotic cells of bacteria had evolved to overcome the problem of a growing shortage of nutritional sources by encapsulating its own 'soup' within cell walls. The early virus probably learned how to attach itself to this new cell. It could then make use of the cell's own productive facilities by penetrating into the rna and the dna to substitute its own coded instructions and so reproduce itself.

It proved something else, too. Mutation arises from dna replications going wrong. Curiously the forces of Nature that determined so early the behaviour of organic descendants seemed to lead, as often as not, to ever-greater degrees of degeneration. This ultimately spelled death, rather than progress, for many species, in spite of recent controversial research suggesting that some bacteria can engineer their own useful mutations (*see* Chapter 1). The blind movement of molecules still threatens existing species with extinction, as when a new dna strand is formed to create an even newer enzyme that ultimately gets replicated into future organisms. Very occasionally, it turns out, this new organism is the product of an enzyme with one faulty amino acid. Herdsmen and farmers have routinely taken advantage of farmyard species that developed mutations (for example, by breeding a variety of cow to yield more than average milk), by selecting from each generation that which seemed suitable for exploitation.[1]

But most mutated offspring are incapable of developing, or instead they die young. Because there are so many genes in living organisms, with billions of replications going on, possibly as many as 40 per cent of creatures are carrying some mutations, though often the worst effects are mitigated because another healthy matching gene 'covers up' or 'helps out'. Perhaps a tissue stays healthy when the myriad other cells crowd out the faulty one. Dna sensitivity can be changed by a variety of biochemical and physical techniques dependent on the growth state of cells, or when they freeze or dry out, for example. Generally speaking, it is the ever-onward march of organic growth that cures most ills, as in the healing process itself. Cells grow and divide to replace those that are lost or damaged. Even when a less perfect or efficient species is produced cells are overrun by more perfectly designed species. Occasionally a mutation is highly beneficial — as in the classic case of the very first giraffe born with a slightly longer neck than his peers. This is likely to have happened with the change-over to a different type of plant, or with the death of unicellular ocean-going creatures, or even with a certain type of dinosaur. On the other hand, cancer often occurs when the growth cannot be stopped at the right time, when it multiplies helplessly.

Events in outer space can play both a malign and a benign role. Sub-atomic particles can smash right through the nuclei of cells housing chromosomes — something chemicals by themselves cannot reach — so damaging the atoms which make up the molecules. The first negative effects of UV were achieved in 1801 by J.W. Ritter, who was checking the way different wavelengths could blacken silver chloride. He noticed that the darker region beyond 380 nm (the violet end) was more effective than visible light. The first discovery of the way UV could affect living systems was made in 1877.[2] The mutagenic effects of radiation were determined by the American biologist Hermann Joseph Muller, in 1926, with experiments of X-rays on flies. A more important discovery was made two years later when the relative effectiveness of different wavelengths of the radiation that could destroy bacteria was seen to parallel the absorption spectrum of the nucleic acid bases. Then great strides were made in the science of photochemistry.

Nowadays much more is known about the harm radiation can cause. According to Claude C. Albritton Jr, experiments at the Brookhaven National Laboratory in New York show that forests can be damaged by exposure to levels of radiation on a par with those nuclear explosions that affect humans. Pine trees are especially sensitive; 2 rads of radiation per day endured for six months can severely affect them. They could be killed outright by six months exposure to 20 to 30 rads per day. Weeds are more hardy, surviving up to several thousand rads, while mosses and lichens can survive up to 200,000 rads.[3] The Brookhaven experiments showed that oak and pine forests suffered a progressive decline after exposure to gamma rays for several years.

The implications of all this have not been lost on the health administrators of modern societies. A doubling of present incidences of active UV would probably be disastrous for the global ecology. The Environmental Protection Agency (EPA) of America predicts a decrease of one percent in density of the ozone layer would yield an increase of 2 per cent in UV at the surface. Scientists say that UV-B rays could easily weaken the body's immune system to infections that enter by the skin, as with bilharzia, the parasitic infection common to the Third World. According to Bob Whitten and Sheo Prasod, who work, respectively, for the NASA Ames Research Centre and the Jet Propulsion Laboratory, California, two types of cancer can be caused to humans: localized ones and the more serious melanoma, which is often fatal.[4] A 1980 EPA Study said that in the United States a one-percent decrease in ozone would result in approximately a 4 per cent increase in non-melanoma cancer.

There is evidence of squamous-cell cancers occurring on the ears and noses of Australian sheep, more so than usually found in temperate areas.[5] It seems that reptiles, birds and plants can survive larger doses than mammals of equal bodyweight. Indeed, an organism's vulnerability to radiation appears to be closely related to the size of its chromosomes. The larger plants, with larger chromosomes, are less vulnerable.

Death in the Seas

Most scientists accept that the UV-B dose can penetrate below the surface of water, and hence poses a threat to organic life there. They also believe that harmful radiation played a role in the extinction of simple-celled creatures many millions of years ago. And they know how and why this could have happened. Laboratory tests, for example, with mercury lamps, have proved that the aquatic organism known as *Phormidium* was found in experiments to stop moving after just thirty-five minutes, even though other parts of the solar spectrum were increased. Filamentous cyanobacteria live in both aquatic and terrestrial ecosystems, and can be killed off within a few days of quite a low dose of UV-B.[6]

There is a general agreement that a great deal of aquatic organic sea life, such as plankton, ammonites and belemnites, did suffer severely at the end of the Mesozoic and that this was largely caused by radiation. The skeletal remains of marine calcareous plankton (nannoplankton and planktonic foraminifera) make up half the total of chalk formation at the top of the Mesozoic strata in the Upper Maastrichtian (the last few million years of the Cretaceous period). They consist of countless millions of nannofossils averaging less than 10 microns in a few cubic cm of chalk. These species all seem to have died out in less than a million years, and probably in no less than many thousands of years.

The implications of harmful solar radiation for the wellbeing of organic life on Earth are profound. Micro-organisms like plankton play an important role in both aquatic and terrestrial systems because they are at the base of the food chain. They seek out niches for survival, and look to signals in the micro-environment to help them do this. Light is naturally the most important energy source, not just for photosynthesizing organisms but for other varieties, too.[7] Many micro-organisms, like procaryotes, and many of the larger gliding and flagellated forms, are apparently already under stress without any extra UV-B. The larvae of fish, as well as crustaceans and molluscs, are constantly threatened. In the modern world the growth rate of many other larger aquatic mammals like penguins are also easily harmed by increases in UV rays (commercial fish and krill harvests are regularly put at risk).

But even microbiological creatures, or at least a good number of them, have built-in biochemical responses to help them minimize the harmful effects of UV. *Phototaxis* is a typical technique of movement towards or away from light, and photopic responses occur with sudden changes in light intensity. Much biological activity takes place in the early morning or late evening to avoid the worst excesses of sunlight. The photopic response is apparent in cyanobacteria, which have been observed to glide into a shadowy area after leaving a bright area, to then immediately propel themselves back into the lit area providing it is not too bright. Admittedly the depth of water itself would have provided an adequate protection from lethal cosmic or solar rays. Radiation can penetrate several metres through clear sea water,[8] and the depth of water that any waveband can reach depends largely on the angle of the radiation, the wave pattern on the surface and the amount of suspended material in water. An organism at a depth of five metres in clear water — fresh or salt — will get about 75 per cent of the dose at 300 to 400 nm, 60 per cent at 320 nm, and 50 per cent at 300 nm. The 300-400 nm band penetrates sea water relatively easily, whereas extreme UV, shorter than 200, would not penetrate at all.

Even so, what makes many scientists believe that UV rays are as potentially harmful as they are life-sustaining is the fact that some simpler organisms do not have the ability to detect changes in UV-B irradiance at any depth, so it is not known whether that have secret avoidance resistances.[9] Often they simply cannot escape the rays. At other times they cannot photo-orient themselves properly, so in turn are unable to survive in an unfavourable environment. Torn between too light and too dark a place, they are forced into the substratum, and die for either of two reasons — they get too much radiation or not enough. Much of the immobile biota, including coral reefs, in estuaries and lagoons, have, for long periods in pre-history, all been regularly wiped out.

Plants, UV Rays and Dinosaur Diet

Plants, for quite obvious reasons connected with their rooted immobility, can also suffer from too much solar radiation. Naturally enough the roots and shoot-tips of plants, being buried in the soil, are protected from UV rays, and these tips are very important for continued growth. Differences frequently occur at the membrane level and at the organ level. And different degrees of protection apply to the outermost cell layer of the leaf, the epidermis. These cell layers often contain substances that quite rapidly absorb UV, depending on need.

As there are variations in susceptibility, there is much competition between vegetable communities, especially when they are grown together. Weedy goat-grasses are more repressed than wheat when the two are grown side by side.[10] But wild oats are more favoured by UV-B than wheat when the two are grown together. Other grain crops, soya-beans and corn are not very sensitive by themselves to UV-B, but wheat and rice are.[11]

Even different parts of the plant vary in sensitivity. Laboratory tests have shown that the lower part of an Oxalis leaf shows more vulnerability than the upper side of the leaf. Pigments also play their part. We have seen that the chlorophyll pigment provides its colour, but plants in addition carry a protective pigment to help cope with harmful sunlight. Even single-celled creatures have this. Fungi varieties can be seen with brown and black spores, and some slime molds have a variety of protective pigments. Other plants grow fewer leaves and produce a thicker, reflective, wax cuticle when exposed to small increases in UV, while others can sprout protective hairs. Plants also vary the way they grow, and can, for example, produce smaller leaves where radiation is most intense.[12]

Most terrestrial animals have ways of cutting down on harmful solar radiation. Gastropods have shells, insects have exoskeletons, reptilian species have scales, and birds and mammals have feathers or hair. Many organisms are either underground or nocturnal, such as worms or snails. Many are quite surprisingly adept at minimizing their exposure by twisting and tilting. Modern plants, as we have seen with cereals, also vary widely in their tolerance, even among closely related species.

It goes without saying that 'repair' mechanisms are vitally important for continued evolutionary survival. Any mutant cell that lacked this regenerative mechanism would be killed by quite a small dose of UV-B or UV-C radiation. Plants, of course, cannot go underground, but they can take advantage of the night. In fact most repair occurs either in complete darkness at night, or in the shade, in a way quite different from another process called photo-reactivation, where solar energy itself does the 'repairing'. One way this can happen is by excision repair, sometimes colloquially known as cut-and-patch repair. This is more likely to happen in bacteria and mammalian cells. Another way is by enzyme repair. In this case a thin strand of the dna, consisting of several nucleotides, including what is known as the dimer, is neatly nipped off by an enzyme. The strand is then 'copied' from the dna polymerase information of the complementary strand, synthesized and reattached.

Plants, in fact, have a tricky relationship with sunlight. They live a precarious existence, with repair mechanisms evolving hand-in-glove with the organism itself. If the leaf does not grow so well this will lead to a lowered input of light energy, and thus a vicious circle is set up. On the other hand there is a protein, called the QB, that has a very high turnover in sunlight. As soon as it is broken down it is re-synthesized. This looks at first sight to be a unique protective mechanism that is in fact dependent upon photo-reactivation repair techniques. But this resynthesization soon tails away as the irradiation is increased with strong light. Strong UV can also affect the closing of stomata and thus again inhibit photosynthesis; a valuable negative feedback, for in this case visible light counteracts UV.

The foregoing discussion of plant genetics is highly relevant to the dinosaur mystery, because the kind of plant species and families that exist today also existed about 200 million years ago. This fact alone very much weakens the argument, often advanced, that the arrival of the angiosperm sub-type of plant played a leading role in dinosaur extinction (*see* Chapter 10). It is widely held that these flowering plants first appeared in the early Mesozoic era, and reached their widest distribution in the Cretaceous, some 150 million years or so later. Flora-induced extinction theories become difficult to sustain for the later generation of dinosaurs born into an angiosperm-dominated world, even if it can be held to apply to earlier species.

However, a radical new theory suggests that not only were the flowering plants on Earth some 200 million years earlier than is supposed, but the way plant genetics change is probably the most important determinant of flora forms themselves. In a *Nature* article, William Marten and colleagues at the Max Planck Institute for plant genetics in Cologne suggest the evidence for an earlier blossoming of angiosperms does not necessarily come from sedimentary evidence.[13] Working out what happened to plant life (as opposed to animal life) at the K-T boundary beyond which no more dinosaur fossils can be found, has never been easy. The record of the Earth itself is based on pollen grains rather than whole plant fossils.[14] For example, most early studies of the extinction of lowland plant life at the boundary have been based on the study of collected pollen grains and marine plankton.[15]

We know, however, that from the beginning of the Triassic plant life on land was evolving rapidly, with the terrestrial habitat being dominated by cycads, ferns, conifers and ginkgoes — with ferns diversifying the most. There were also the pteridosperms, the forerunners of modern palm-trees, which reached their greatest development in the Jurassic. But it seems, interestingly, that the diversity of pollen decreased markedly at the K-T transition with a more than 50 per cent turnover in species of pollen and leaves for both gymnosperms and angiosperms. By the time of the lower Cretaceous the number of taxa were probably limited to a few simply built types. Prior to the K-T, from their earliest beginnings onwards, it is now believed that angiosperms were environmentally adaptable, and radiated in much less than 100 million years. A perceived pattern of divergence of nine extant species occurred between types known as monocots and dicots, with the dicot sub-classes dividing into five more. Starting off in the forested uplands, they later radiated to the lowlands. They then became apparent for the first time in the fossil record.[16]

The Max Planck researchers compared small genetic differences in flowering plant family-trees to determine the time of each divergence. They calibrated the rate of change with animal genes, bearing in mind what was known about evolution, and concluded that angiosperms were much older than had been thought. The reason, they said, there were no signs of earlier angiosperms (say, earlier than the Jurassic), is because they were probably confined to highland areas where fossilization, which normally takes place in sediment-carrying rivers and lakes, would be infrequent.[17]

They give added, albeit unwitting, credence to the radiation theory. Let us assume that the ozone layer, for some reason, was getting thinner. The percentage increase in dose is usually 1.5 to 2.5 times the percentage decrease in ozone, depending on latitude, season and 'action spectrum'. With ozone reduction the latitude limit of UV-B tolerance will move to higher latitudes. Presumably animal populations at these higher latitudes will get a higher dosage of UV-B, but it would be at or below their tolerance limit. This dosage would have been more harmful for the early plants because plant genes may not have changed at the same rate as did animal genes, because, in turn, the 'repair' mechanism may not have been fully developed in all species. But it seems that, for the angiosperms finally to get a toehold on Earth, and to develop an adequate 'repair' mechanism, their future and that of the species that fed on them was secure.

Of course, the fact that trees and plants can become extinct through 'genetic erosion' or through failure to correct genetic faults, has been known for some time. The danger today is that Man, with modern agricultural practices and advances in plant breeding, can purposefully narrow the genetic base by cross-breeding lines with valuable characteristics; i.e. those that make more efficient use of fertilizers, and screening out individuals with undesired traits.

But in nature this can happen by chance. The Irish potato famine resulted from genetic uniformity, caused by a new fungal disease. The narrow genetic base meant that the potato had no natural resistance to the disease. In 1970 the United States lost half its maize crop to another fungal disease because, again, varieties of corn had been bred out so that most remaining species had a single disease-prone gene. In these cases genes from other varieties were quickly cross-bred and used to rescue the stricken American varieties.[1 8]

Cosmic Rays and the Dinosaurs

The word 'radiation' is frequently used pejoratively by members of the public, as it conjures up images of nuclear war and nuclear accidents. Press coverage is often given to the occasional (and often harmless) leakages of radioactive gases into the wider environment. And yet 'radiation' is going on continually all around us, and without it life as we know it could not exist.

When an atom decays, beta particles (high speed electrons) are emitted together with gamma rays (high energy photons from the high energy shortwave blue end of the spectrum, beyond UV and X-rays). They operate like light, on a much shorter wavelength, and travel enormous distances. Alpha particles are from the helium nucleus and travel only a few centimetres. Gammas and alphas can split betas (some 2000 times lighter than protons) from atoms, and attach them to other atoms, thus producing positive and negative *ion pairs*. They can easily penetrate matter to release their energy over much longer paths (i.e. the ionization effects are dispersed wider).

Alpha, beta and gamma rays are ones of increasing penetration but decreasing concentration. This means that with the alpha series the organic damage is done by absorption inside organisms rather than passing straight through it, which is the case with X-rays and gamma rays. Alpha particles may be stopped by a dead layer of human skin. Betas can penetrate a couple of centimetres of tissue, as can gamma rays, depending on distance and on the number. Most of these electrically charged particles shower down to Earth from the innermost depths of space, and not just from the Sun, probably from the vibrant radio stars called pulsars. Energetic gamma-rays are probably hitting Earth from a neutron star circulating around another star known as Cygnus X-3, some 30,000 light years away.

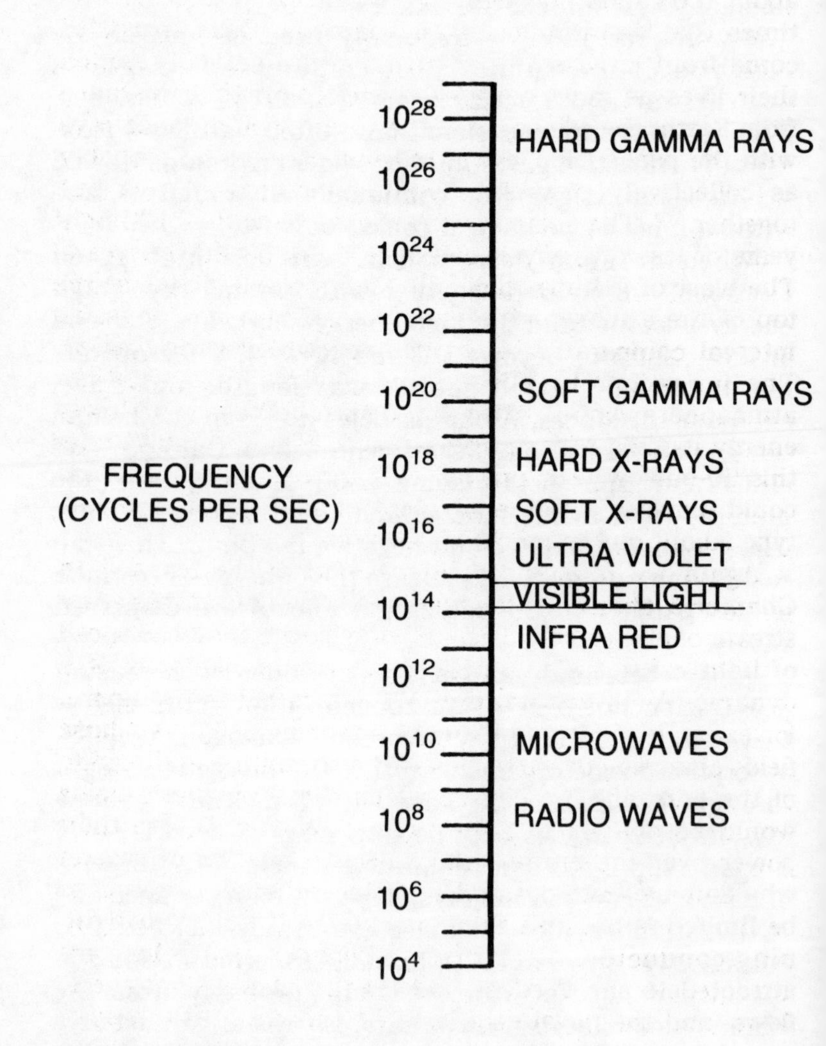

Fig.21. The Electromagnetic Spectrum

However, we must be careful not to be too alarmist about the dangerous-sounding 'cosmic rays'. The intensity of these rays reaching our atmosphere amounts to about 0.03 rads per year. We would need at least 500 times that amount to do any damage. Other particles come from large stars which have dramatically ended their lives as supernovae. Depending on their distance from Earth, the remnants of these stars will produce rays with the penetration power of millions of volts, probably as collectively powerful as all other cosmic rays put together.[19] The solar system should be within 100 light years or less of a supernova once every 50 million years. The blast of gamma rays and X-rays would arrive at the top of the atmosphere over the span of a few days, an interval comparable to that during which it originated. The interaction of cosmic and gamma rays with the atmosphere would yield a powerful shower of lower energy gamma rays at ground level. For a few days after this 50-million year 'one-off' event, background radiation could increase to the equivalent of an Hiroshima bomb-type fallout multiplied 300,000 times.

Again we must bear in mind what we learned in Chapter 5 about Earth's own protective envelopes. Any stream of charged particles moving slower than the speed of light creates a magnetic field, rather like that of a dynamo. A current flowing in a coil round an iron core, for example, can produce very powerful magnets whose fields change with variations in the strength and direction of the current. The curved paths these magnetic fields would be obliged to follow would tend to diffuse their power over the entire galaxy. Hence one major reason why damage from cosmic and solar radiation is bound to be limited is because Earth itself acts like a giant lightning-conductor. Electrically charged particles are attracted to the vertical path of the planet's magnetic flows, and the negatively charged particles of the uppermost layer of the atmosphere. The magnetic lines of force are outwards along longitudinal lines above the Earth. Speeding electrons, protons and other particles from space are drawn to the northern latitudes of these lines of magnetic force, and become trapped.

Even so, the fear is often expressed that Earth's magnetic field could fail from time to time, or change the direction of its polarity, thus allowing both excessive amounts of UV radiation and more cosmic particles to reach the atmosphere. Most of the 'cosmic ray' dinosaur extinction scenarios involve complicated magnetic field reversals over periods of tens, thousands or hundreds of thousands of years. Sometimes the field is said to reverse itself every 220,000 years,[20] with the reversals taking between 20,000 and 50,000 years to complete the cycle. Some astrophysicists suggest that during the last 2.5 million years eight species of unicellular marine creatures became extinct after magnetic reversals, with the organisms becoming disoriented when the magnetic grains in their bodies became disturbed.[21] The distinguished astrophysicist Bill McCrea once pointed out that both the Permian and the K-T extinction occurred during episodes of frequent magnetic reversals.[22] Dale Russell, a paleontologist with the National Museum Board of Canada, also became well known for his assertion that a magnetic 'flip' could have terminated the lives of the dinosaurs.

Frank Close, a leading physicist at the University of Tennessee, and author of the popular work *End: Cosmic Catastrophe*, adopts a more negative approach and says there was evidence of two supernovae, leaving pulsar remnants 456 and 196 light years away, that did cause an increase in cosmic radiation in our vicinity 62,000 and 440,000 years ago. This then reacted with the planet's atmospheric nitrogen and produced abnormally large amounts of radioactive carbon-14.[23]

Again some scientists, such as Donald Goldsmith of the University of California, deny that a disappearance of Earth's magnetism would be harmful, saying that damaging radiation at the surface of the Earth would only rise by 10 per cent more than is usual, which would be much too low to kill off entire species. Any biotic stress that could arise from cosmic rays in the upper atmosphere would normally be dumped as heat into a layer extending to the stratosphere and mesosphere, causing turbulence and heat-retaining properties of the atmosphere, which in turn would cause storms. It would circulate low, water-laden air into higher, drier levels. There it would freeze into high-altitude ice clouds, and this could bring on a mini or even a full-blown Ice Age.

In Chapter 8 we will return to the subject of cosmic events bringing about changes in Earth's climate.

Did the Dinosaurs go Blind?

In the meantime it is worth introducing an original and intriguing theory about the death of the dinosaurs that received scant attention when it first appeared in print in 1982. We return once more to solar radiation. R. Croft is a biochemist who advanced a theory that the surviving dinosaurs, under a Sun he assumed was growing more harsh, eventually developed cataracts in their eyes, ultimately going blind and losing out in the survival race.

His theory has some credibility because we know that the eyes of a wide class of living creatures can suffer in this way.* It is worth reminding ourselves that cataracts in Man are indeed more common in the hotter regions of the Earth. A one-percent depletion of the ozone layer can increase cataracts by between 24,000 and 57,000 cases a year.[25] There is also, curiously, proof of incidences of cataracts among eskimo races, evidence that it is largely the UV spectrum that is to blame, since eskimos suffer more acutely than other races from the dazzling effects of sunlight being bounced off white ice sheets.

Of more relevance to our argument is what can happen to the animal eye under fierce sunlight. Although wild

* Incidentally, *Moscow News* reported in 1984 that dinosaur fossils showed an unusually high uranium content, and may have been killed off by 'radiation' in the lagoons where they lived.[24]

animals generally are not very sensitive to enhanced UV-B, some domesticated cattle, like the white-faced Hereford cattle, *are* susceptible to cancer of the eye. UV-B varies with latitude, as we have seen, although there is little evidence of latitude-dependence effects other than for the eye.

Croft's evidence that blindness afflicted 'the last dinosaurs' (the title of his short book)[26] is the famous coal-mine shaft near Mons, Belgium, where a vast number of dinosaur fossils were found entombed in a rock fissure in the mine. This he took to be a lethal trap for the unwary, with entire herds of iguanodons falling into the ravine. He points out that no young were found in the fossil remains, leading to the conclusion that the cataract blindness was a gradual disease, afflicting only the older members.

The biochemical structure of the vertebrate eye has remained constant throughout evolution, so Croft assumes the dinosaur eye was similar to that of many other species alive at the time. Croft cites the work of George Walls, a professor of optics at the University of California, who claimed that certain evolutionary lizards were forced to adopt a subterranean existence during which time their eyes degenerated, and they became a form of snake. He also cites the case of vestigial eyes in fish accustomed to abyssal depths of the ocean.

Solar radiation, said Croft, allowed only those with brow crests to survive. The post-orbital bone overgrowth of some dinosaurs was a form of sun visor — conferring immense evolutionary advantages at a time of rapid solar change. The hadrosaurs and the proceneosaurs also had cranial crests. In each case they jutted out just above each eye. Some were hollow and others had a solid crested mantel. In *Saurolophus*, discovered in Asia, the crest was found to be made from an enlarged nasal bone. *Pachycephalosaurus* had a massively thick skull with elaborate bony knobs on the snout, the rear of the skull and over each eye. The bony frill in *Torosaurus*, said by other scientists to correlate with the enlargement of the jaw muscles, also seemed to be a protection from direct solar radiation. Were they battering rams, as has been suggested? If they were, says Croft, why did the thick bone develop over the eye first? *Ceratosaurus* had a very thin skull, yet still had bony projections in front of each eye; and its close relative, *Dilophosaurus*, also had a pair of bony plates situated above each eye. *Allosaurus* and *Tyrannosaurus* also had bone thickening or projections over the eyes, the latter, according to John McLoughlin, protected from glare by a 'supra-orbital ridge like that in eagles.'[27] Brow crests attained their maximum development in *Triceratops*, especially the sub-order enryceph-alus, with its massive brow horn-cores. The genus cera-tops was originally named by Othniel Marsh in 1888, from the Greek words meaning 'horned eye'.

Croft points to the possibility of the surviving archo-saurs in the Cenozoic being nocturnal, and mentions reptiles like the tuatara. This creature can control its exposure to sunlight by means of its media eye, some-times referred to as a 'third eye' or 'pineal eye', which acts as a kind of dosimeter for solar radiation. In exper-iments it was found that the American chameleon (Anolis carolinensis), when it had its pineal eye covered with a piece of aluminium foil, was unable to control its expo-sure to sunlight. The dinosaurs, claims Croft, did not have this third eye, and so were more susceptible to overheating.

Scientists agree that desert iguanas can see ultraviolet light, following experiments done by Allison Alberts of the University of California at San Diego.[28] The scent marks they leave to demarcate territory and pass on messages to other members look faintly green and fluor-esce under UV light. Alberts suggests that desert buds and flowers, often visually inconspicuous, absorb UV light strongly.

Croft's thesis, inevitably, has been challenged. Writing in the *Freethinker* in April 1982, T. Soliar said that Croft had a poor understanding of dinosaur anatomy, since the eyes of many of them were invariably situated towards the sides of the head, with the horns facing forward. There is another problem. The protein in the lens of dinosaurs must have been similar to modern vertebrates, because proteins seem to have changed little during evolution. Some evolutionary biologists believe the dinosaurs had colour vision in the same way (as far as is known) reptiles do. It has been suggested, for example, that the emergence of the bright flowers in the Jurassic acted as a warning.[29] Furthermore, we have no way of knowing whether dinosaurs were entirely daytime creatures, and neither are we certain that they did not possess many of the other anatomical features pertaining to modern-day reptiles, none of which has so far shown signs of cataracts. Many hadrosaurs had huge eyes, with eyeball diameters of about 10 cm, possibly implying they could see at night as well as during the day.

The photosensitive part of the vertebrate eye, the retina, is composed of cells in the form of rods or cones containing pigment. There are two types of pigment; one for night (rhodopsin) and the other for daylight (iodopsin). The weakest light pulses which allow the eye to see send about 100 photons through the pupil via the iodopsin pigment. When the pulses become weaker they are drowned out by heat-induced random rhodopsin reactions which are very similar to changes triggered by light. Chemical changes in rhodopsin molecules can also be triggered by light pulses, which eventually send signals to the brain.

Here we can speculate whether the Mesozoic climate went through one of its periodic fluctuations (*see* Chapter 9), when solar radiation on both UV and IR wavelengths was dramatically increased over an evolutionary space of several tens of millions of years. If this was so, then the knowledge we have gleaned from a variety of zoological and biochemical research sources could prove that the dinosaurs were cold-blooded after all. There are a number of ways this can be worked out. Firstly, while rhodopsin molecules triggered by a given amount of light increase exponentially with eye size, thermal noise (i.e. a general background rise in temperatures) only increases in a linear fashion, so that bigger eyes are more sensitive to low light levels. But temperature itself has a different effect on the diopsin pigment, and could affect lens transparency. In *Nature*, in August 1988, an article appeared concerning the eyes of cold-blooded creatures and the quality of vision under changing ambient temperatures.[30] It appears that in normal circumstances the eye lens is maintained at a temperature slightly below that of the rest of the body, and it is easily susceptible to thermal damage. Lens thermoregulation in birds is controlled by the pecten, which is a packed series of blood vessels arranged in a configuration like a vane, although it is far less developed in nocturnal birds. Ann-Christine Aho of Helsinki University reports that laboratory researchers found that toads could see their prey in light eight-times dimmer than would be perceptible to the human eye.

Here the thorny question of metabolism becomes relevant, because cold-blooded animals are less able to regulate the temperature of the eye. Some snakes and lizards have pectens, but we cannot know whether dinosaurs also had them. Some hadrosaurs, we know, had a space behind the eye where there was a large blood sinus. Could this have been used to lower the temperature of the eye? It seems unlikely, because R. Croft and his biochemist colleague M.B. Tabet did an experiment with purified proteins exposed to sunlight, and found that the thermal and light-stable protein is only present in warm-blooded animals. Hence development of endothermy in mammals from ectothermic ancestors involved the acquisition of vital thermoregulatory mechanisms. It seems, then, that dinosaurs' lens must have lacked the stable protein and *were* hence susceptible to cataract blindness in the event of abnormal solar activity, and they would have been, at least according to the logic of Croft and his colleagues, 'cold-blooded'.

Chapter 7
THE DEATH STAR

THE IDEA that the dinosaurs were killed off by a missile from outer space, recurring in many magazine articles and popular scientific works since the war, moved centre stage in the 1980s. The first real academic pronouncement on the extraterrestrial (ET) hypothesis was made in 1979 at the American Geophysical Union meeting in Washington. And from 1980 onwards — starting with an article in *Science* magazine — the argument has been conducted largely in learned journals, with the world's press eavesdropping, and reporting, on the theory with growing interest.

Somewhat to the surprise of many scientific bystanders in other disciplines it has been a sustained argument, and with every year that passes without conclusive and proven rebuttal the case for a catastrophic end to the dinosaurs, as well as other species in other epochs, gains credibility. It has, in other words, been removed from the domain of the pseudo-scientist and given academic respectability. Indeed the academic pendulum, if not exactly swinging, is inching these days decidedly towards catastrophism. Specialists from most of the earth sciences, as well as many astrophysicists, are slowly (some reluctantly) reshaping their ideas. A recent book of scientific readings on the subject of mass extinctions brought together the discoveries and explanations of many experts from diverse disciplines. And yet the (hitherto orthodox) non-impact theories of extinction were hardly given a mention.[1]

Yet, as we shall see, there is still a need for a healthy scepticism about the ET perspective, and for greater balance to be given by commentators. The ET viewpoint may now have equal status with the geophysical one, but this does not yet mean its proponents have won the argument, since the geochemical evidence upon which the argument is based is open to other interpretations.

Briefly, the causative factor in the extinction scenario is said to be the massive pall of dust which was ejected into the atmosphere after some kind of missile landed, and which blotted out the Sun's rays for months or, as was originally suggested by a leading team of researchers, for years, to suppress photosynthesis. It was this final event that brought about the collapse of the food chain. Other variations on this theme suggest instead that a vast tract of forest caught fire, and smoke asphyxiated many creatures, or instead the chemical balance of atmospheric gases was somehow lethally disturbed to produce sulphuric acid aerosols.

It is curious to note that the ET theory of dinosaur extinction was inadvertently arrived at by an eminent scientist searching for something quite different. At the medieval town of Gubbio, situated in the Umbrian Apennines of north-central Italy, there is a prominent layer of reddish-grey limestone about 2 cm thick, above which is another layer of clay packed with fossils. These sediments have for years attracted earth scientists from all over the globe as it is the dividing line between the Cretaceous and the Tertiary periods (the well-known K-T boundary). In the past the pink limestone has been used as a desirable building material, formed from deep-sea sediment laid down over 155 million years ago from the Jurassic to the Oligocene.

Sandwiched within the clay boundaries is to be found a strange metallic substance now enshrined in geologic history: *iridium*. Iridium comprises approximately one 10-billionths of the total material of Earth's crust, but at the K-T boundary it is up to thirty times that normally found in other terrestrial rock. Iridium is element 77 and belongs to that group of metals known as 'platinoid', so named because the most famous member within the group is platinum. It is a very hard substance, like its close neighbour osmium (element 76), and very heavy. In fact, platinum jewellery contains about 10 per cent iridium, added to make it harder. There are other iridium sites — in New Mexico at the Baton Raton basin, and at a Danish site known as Stevn's Klint. There is also a hint of iridium at the Cenomanian-Turonian boundary, about a million years prior to the K-T boundary. A Chinese group has reported iridium at the Permian-Triassic boundary.

The quest for the explanation for species extinction centres around this metallic layer found in the K-T strata (known as an 'anomaly'), since it may naturally enough have a bearing on why no dinosaur fossils can be found *above* the anomaly (although this fact, concerning the stratigraphic position of dinosaur fossils, is sometimes refuted).

The distinguished late physicist Luis Alvarez, formerly of the University of California at Berkeley and a Nobel Prize winner, had been examining stratigraphic evidence for years, and had been interpreting it within a conventional paleontological frame of reference. Earth scientists claim that the iridium was outgassed from volcanoes, along with other dust and debris, or perhaps other terrestrial processes were involved as the Earth underwent physical change. Indeed a scholarly knowledge of what gives rise to earth substances excludes any ET explanation almost by default. There are natural and recognizable processes by which earth elements are distributed and concentrated in certain areas. Rock strata are continually deformed, heated and cooled, to become widely dissipated in surrounding materials.[2] There are biological processes, too, such as bacterial decomposition, which can separate out metallic compounds, and sedimentary material may be sharply divided by compression so that further blending of soft material ceases, and so on.

However, the knowledge and imagination of the astronomer, especially the planetary scientist, would be equal to the challenge, as we will soon see, since he knows that the origin of *all* solid objects in space, whether planets, moons or asteroids, have an identical source. The argument is, and probably will remain for some time, a two-way one, because ET events, in theory at least, can still alter the chemical composition of the physical evidence. In other words the evidence itself could very well mislead the serious researcher regardless of how rigorous his methods.

Iridium From Volcanoes?

It was Charles B. Officer and Charles I. Drake of Dartmouth College, New Hampshire, who advanced the theory that the iridium comes from deep within the Earth. In their 1985 articles in *Science* they said they had calculated that the layer was deposited over a period of between 10,000 and 100,000 years, and that the mechanism for its deposition was continental drift reaching a decisive phase which triggered volcanism into life.[3] In fact tremendous kinetic heat must have been needed to melt rock and to spew the spherules (tiny pellet-like objects showing signs of having been forged under great heat and pressure) — many containing the incriminating iridium — across vast areas of the Earth's surface. This, to many theoreticians, was proof enough of volcanism, as Jan Smit of the Geological Institute of the University of Amsterdam pointed out, after he found spherules in clay samples from Spain.

The volcanists cite the Deccan Traps in India as evidence of violent volcanic activity in the distant past at a time of great geophysical turbulence, as continental drift forced India to join up with the rest of Asia. The Deccan Traps are to be found in the Western Ghats mountains south-east of Bombay, and are believed to be the product of the most extensive lava flows anywhere in the world. They are known to have been deposited over a period of less than 900,000 years, anywhere between 69 million and 65 million years ago. They cover a phenomenally extensive area: more than half a million cubic kilometres. The Deccan volcanoes would have forced enormous fountains of fiery lava and poisonous gases and ashes into the atmosphere, which would have made conditions deadly for many species of plant and animals.

There are other incidents recorded by the Earth itself of violent volcanism. The volcano Toba, which erupted about 75,000 years ago, blasted 400 times more debris into the stratosphere than even Krakatoa did. Some 30 million years ago, there was repeated volcanic activity in what is now known as the Far East. There are also many square miles of petrified forest around the Yellowstone Park area of Wyoming, with successive forest layers buried in a kind of rock formed from volcanic ash.[4]

It was at this stage — discovering the iridium and working out a grand theme to account for it — that the cosmic angle emerged. In the early 1980s there appeared to some scientists evidence that extinctions on Earth had been cyclical — recurring time and again. But there was no known periodicity, or so it was thought, inherent in any natural Earth mechanism.

In the meantime other scientists were demonstrating that iridium is essentially an alien substance, to be found in considerable amounts in comets and meteorites. A major proof for the ET argument was said to be the gold and platinum elements along with the iridium, which differs considerably from what could be found in Earth's crust. Taking into account the quantity of discovered iridium, the Alvarez team calculated that the clay layer hinted the crater was 100 miles across, and data for impact craters also suggested objects must be a minimum of five miles across to make this kind of dent.

There was little agreement over the meaning of the spherules found along with the iridium. To confuse matters the larger microtektites (probably oceanic-terrestrial) and the smaller particles (probably of cosmic origin) were found at Gubbio. R. Ganapathy, a geologist working for a chemical company in New Jersey, believed that metallic sediments were virtually identical to those of metallic meteorites, and even the stony spherules were similar to those discovered in stony meteorites.[6] Richard Muller, a physicist who worked closely with Luis Alvarez, said that about 200 grams a day of iridium hit the Earth from approximately 400 tons of meteorites landing on the surface.[7] Alvarez himself had calculated that 50,000 tons of the iridium had been deposited at the K-T boundary. So assuming only a fifth of this remained from the pulverized object, said Muller, one could say with reasonable certainty that it contained 250,000 tons of iridium, making a mass total of all other substances combined of 500 billion tons. And, according to other astronomical calculations, it would have been at least eight kilometres across, or about five miles.[8]

The volcanic thesis soon, as a result, had a major contender. Indeed Frank Asaro, a nuclear chemist at the Berkeley-Lawrence Laboratory in California, said he could identify volcanic layers by the *absence* of iridium,[9] and not even those powerful volcanoes like Kilauea showed any excess of iridium. Karl Turekian, a geochemist at Yale, was the most prominent of the pro-volcanist persuasion, but his conversion to the ET school of thought at last convinced many sceptics.[10]

In fact the presence of other cosmic dust found along with the iridium layer led Luis Alvarez to change tack. Originally he wanted to check out the clay at the boundary to see if there were any reversals in Earth's magnetic field at Gubbio, and so develop a more elaborate extinction theory. So he switched experiments to determine the timelapse involved, bearing in mind micrometeoritic materials reach the surface of the Earth at a steady rate.

Nevertheless the dispute between the geophysicists and the astrophysicists centred around the crucial need for evidence which would be more than circumstantial. Iridium is a very rare Earth substance, but the asteroidal deposit of iridium would be equally as rare and could also, like the volcanic thesis, be proven only circumstantially. Many believe that a bolide-like object (a missile violently reduced to plasma by the heat of entry) would actually have dispersed even more iridium than the few centimetres so far discovered. Furthermore, some scientists disagree with Frank Asaro's assertion about the absence of volcanic iridium. William H. Zoller and his colleagues at the University of Maryland at College Park, recently *did* discover unusual concentrations of the substance in particles emitted by the Kilauea volcano in Hawaii. French earth scientists have also found similar particles from a volcano on the island of Reunion.[11]

The evidence seems confused, however, about the meaning of the 'shocked' quartz grains — a mineral, like the spherules, subject to violent heating and compression. Bruce Bohor, a clay mineralogist, and his colleagues from the US Geological Survey at Denver, reported that the presence of minute crystals of quartz in the clay layer was proof that it *was* subject to severe shock. Compare this with the evidence of recent research performed by scientists at the Lunar and Planetary Institute at Houston who found no shocked minerals in volcanic rocks at the K-T anomaly.[12] And those scientists who understand the difference say that known volcanic spherules appear to be quite different from those found along with the K-T iridium layer. In any event, it was maintained, seismic or volcanic activity would yield and contort rather than shock or compress. And, according to other experts, the iridium layer was found to be too far up the Earth's mantle to give much credence to the volcanism theory. Furthermore, the microtektites prove that the object did not disintegrate completely.

Other scientists like Miriam Kastner, a geochemist at the Scripps Oceanic Institute, said the clay layer in Denmark and in the Pacific was made up largely of *smectite*, a glass-like compound not known to come from volcanoes.[13] At the Danish site the boundary clay was found to be enriched with platinum, as well as palladium, ruthenium and rhenium — all very rare minerals.[14] And yet many of these were all found well above the boundary clay in deep-sea sediments. A research ship named the *Glomar Challenger* found iridium embedded in cores of clay dragged up from the ocean floor, and revealed oceanic basalt-type sediments. This suggests, incidentally, that if any meteorite had caused the extinctions it must have landed in the sea. To Kenneth Hsu, of the Swiss Federal Institute of Technology at Zurich (and who favoured a comet as the main cause of extinction),[15] the metal was not only the result of cosmic fall-out but the logical settling down of it as it reacted with sea water.[16]

The dividing line between the geophysical and ET explanations becomes blurred because of the way the solar system was first created. All the solid members, as we saw in Chapter 1, were derived from the same atomic materials. Iron, as well as silicates, carbon and other particles were spewed out from the exploding 'Birth Star', but when the Earth cooled the iron elements sank to the centre. In meteorites, however, they remained randomly distributed throughout the object, because gravity within it was not strong enough to draw the iridium inwards. This is why there is more of the metallic substances in meteorites than in the upper layers of the Earth — unless they were somehow blasted there from volcanoes, according to the alternative argument.

In fact an intriguing line of reasoning suggests that the iridium, like the other metals, got ejected from a supernova at times after the Earth was fully formed. If this was so it would considerably strengthen Croft's belief in eye cancer arising from a radioactively poisoned Earth (*see* Chapter 6). The problem with this theory is that there is an inadequate consensus of the kind of particles thrown out from them. Be-10 is an isotope of beryllium with a total of ten protons and neutrons created by cosmic radiation when the atoms in the atmosphere split apart. This happens continually, and provides a useful yardstick to measure the passage of time. Belief in the supernovae theory is also derived from the knowledge of iridium's two isotopes — 191 and 193. This isotope ratio would be the same throughout the universe in particles emanating from a star or from a meteorite.

Others dispute this, saying that iridium 191 and 193

are not to be found in stars. Iridium was instead thought to be a solar system substance — not known to be found in stars, either, as the typical condensing-out material. There was also insufficient of the other major telltale supernovae ingredient — plutonium 244. In any event, over a period of time the 'signature' of particles would differ slightly, so it would be difficult to say precisely what ingredients a supernova should be blasting out. Other scientists pointed to the improbability of a massive supernova explosion. Helen Michel, for example, another nuclear chemist, also working with Frank Asaro at the Berkeley-Lawrence Laboratory, did neutron-activation analysis on the iridium layer to discover the isotopic ratio. She found that it was so high an exploding star would have to have been a mere tenth of a light year away — hardly more than a month's distance at the speed of light.[17] In other words, there would have been a mere one in a billion chance of one occurring so close in the last million years.

But if the supernovae explanation for iridium could be discounted, another ingenious method could be used to test the ET theory. Chemist Jeffrey Boda of the Scripps Institute and his colleague Meixum Zhao decided to search for ET amino acids. At least 55 different acids have been found in meteorites, and only twenty of those are used by living organisms on Earth to build proteins. Tests around the Stevn's Klint site paid off, as Boda and Zhao succeeded in isolating two different amino acids whose origin was almost certainly extraterrestrial: alpha-amino-isolutyric acid and isovaline. Indeed the quantities of amino acids were one hundred times more than those of iridium.[18]

Do Extinctions Occur Episodically?

The debate about iridium took on a more profound implication when it appeared, often to the consternation of paleontologists and evolutionary biologists, that extinctions seem to have occurred in a periodic or cyclical manner. Richard Muller and Walter Alvarez, the geologist son of Luis and also at the University of Berkeley in California, concluded, from studying the better-chronicled craters dating back 250 million years, that up to 95 per cent of life on Earth had been wiped out on at least three occasions (the Great Permian was a notable example). On seven other occasions, they said, up to half of all species were eliminated — with most of them seeming to agree with the new theory discussed in scientific circles that extinctions seemed to occur at 26-million-year intervals.

Erle Kauffman, a paleontologist at the University of Colorado using radiometric and chemical methods, found with some surprise four mini mass extinctions during the Cretaceous itself. Each extinction occurred over a period of time of less than a million years between each,[19] as well as other extinctions at the end of the Triassic.

The 26-million-year extinction cycle had been suggested by paleontologists at Chicago University. John Sepkoski Jr began working backwards over 350 million years, with the computerized aid of a working list of 3500 separate fossil taxa often used by paleontologists in their research. He concluded that as many as 3500 families of marine animals, perhaps covering as many as 250,000 different species, had become extinct. It was then that Sepkoski, assisted by David Raup, came to the conclusion that major extinctions seemed to have occurred at 91, 66, 37 and 11 million years ago. They found, in other words, a rough 26-million-year cycle. The first was at the end of the Permian with some 90 per cent of marine species disappearing.[20] At the same time a pattern of cyclical extinctions over a long time-cycle of 32 million years seemed to emerge from the research carried out by geologists Alfred Fischer of Kansas University, and Michael Arthur of Princeton. However, they soon agreed with Raup and Sepkoski's findings, and revised their 32-million-year figure down to 26 million years after taking into account the age of rock strata.

Now the ET argument was beginning to take on a new and more exciting dimension. What could cause periodic extinctions on Earth? The answer — missiles from outer space. The evidence comes from the observed surface of the solar planets. Callisto, one of Jupiter's moons, is 3000 miles across but has its entire icy surface pitted and cratered. The *Voyager* spacecraft, on its visit to the outer planets, also showed the outer moons to have impacts. Mercury was nearly destroyed by a celestial missile.

It is known, too, that 500 meteorites hit the Earth's surface every year. These are the small ones that have survived the journey to the surface. Smaller objects less than about 150 metres across would likely be broken up by the kinetic forces of entry into the atmosphere. A 50-ton asteroidal object could arrive every thirty years, a 250-ton one every 150 years, and a 50,000-ton behemoth every 100,000 years. Another class of missile, about half a mile wide, could land once in a quarter-million years. This is what astronomers can calculate as a statistical average. This average, in turn, is arrived at from an examination of over 100 terrestrial craters that have been discovered.

The Manson crater, in Iowa, spans ten kilometres and could be the likely site of the K-T impact; but scientists say this is too small. Another series of craters — the Kara craters — in the Soviet Union are also likely candidates, but are thought to be about ten million years too old.[21] Observational astronomers, using new cameras able to operate at night, can spot 1000 boulders in the night skies on an annual basis — some up to a mile wide.

Then there is the evidence of near misses. A 4-metre meteor weighing an estimated 1000 tons was seen by startled eye-witnesses in the western United States in 1972, and two years later a 200-ton fireball was seen streaking through the skies of Europe. These are known as Earth-grazers. One nearly a mile wide was also spotted in February 1982.

One of the most famous depressions is the meteor crater (1.2 km across), sometimes known as the Barringer crater after the name of the scientist who first discovered it, in Arizona. It was probably formed some 50,000 years ago, creating a rim rising nearly 60 metres above the surrounding countryside. Scientists have estimated the weight of the object that caused the impression to be anywhere between 12,000 tons and 1.2 million tons, and from 76 to 366 metres in diameter.[22] The largest known meteorite on display is at the Hayden Planetarium in New York, and weighs about 34 tons. Moreover the periodicity theory is derived from the fact that scientists at the Goddard Space Institute have found that over forty Earth craters formed over the past 250 million years were clustered at 30-million-year intervals, a reasonably close approximation of Raup and Sepkoski's figure. The astronomer David Hughes, in a 1979 *Nature* article, said that one-kilometre craters are created every 1400 years; a ten-kilometre one every 140,000 years and 100-kilometre ones every 14 million years.

So Muller and Walter Alvarez, who worked closely with his father for several years on the dinosaur extinction mystery, came up with another similar figure for just thirteen craters — 28.4 million years. A brief debate ensued concerning what kind of celestial missile could be deflected to Earth at regular intervals. All the early theories suggested it was a comet, but later more substantial objects like asteroids were favoured.

Asteroids are larger and much, much fewer in number than both meteors and comets: according to some estimates no more than 50,000 of them, probably even fewer. They are, however, generally massive — more like miniature worlds known as 'planetesimals' or even 'planetoids'. They lie in more secure orbits in the asteroid belt between Mars and Jupiter. Some of them have more elongated ellipses, however, being pulled out of line either towards the Sun or towards Jupiter. Others are locked into Jupiter's orbit.

The worst threat to Earth is when some stray too far beyond Mars in the direction of the Sun towards Earth. The German astronomer Gustav Witt discovered such an asteroid in 1898, which became known as Eros. It had a diameter of fifteen miles, coming as close to Earth as the Sun now is. It has the potential — when at the right point in its orbit — to get as close as 14 million miles. Since then many other 'earth grazers' have been discovered, like Hermes in November 1937, which was spotted streaking past Earth only 400,000 miles away — a veritable near-miss in celestial terms. It was on a trajectory taking it easily closer to us than is our moon.[23] Earth had near-misses with giant asteroids in 1976, 1982 and again in 1987. Our narrowest escape to date occurred in March 1989, when a giant asteroid approached the planet at a distance equal to twice that between Earth and the moon. It was the closest approach of such an object since Hermes was spotted. Astronomers at the University of Northern Arizona calculated that the object, travelling at 46,000 mph, orbits the Sun once a year on an elliptical path that regularly pulls it back to Earth.[24]

Especially massive asteroids like Hermes, a mile across, are known as apollo missiles because the first of these, named Apollo, was discovered in 1932. The larger apollos like Ceres, 600 miles across, are really solar moons. Ceres is about the size of France and the Low Countries combined. Juno, another massive asteroid, is about the size of Ireland, and Vesta would cover southern Scandinavia. The apollo objects are quite numerous, winging out of the asteroid belt and probably no more than three miles in diameter on average. But they would all have potentially catastrophic results should they strike Earth. Eugene M. Shoemaker, an astro-geologist with the US Geological Survey, estimated the existence of at least 700 apollo-class asteroids over one kilometres in width. He reckons a 10-kilometre wide apollo could collide with Earth once every 100 million years.[25] Some are spotted in the sky and then 'lost', which means that we are not always aware of their trajectory and schedule. Some 'die' and get broken up to become part of the annual meteor showers, but because they can whirl in and then streak away they can easily cross Earth's path and collide.

The Menace of the Comets

Comets are a markedly different kind of cosmic object. A comet consists of an icy, partly solid nucleus trailing a massively long coma. This is the gaseous tail of the comet which can be as much as 100,000 miles across, with the solar wind always making it point away from the Sun. The nucleus probably contains deep-frozen ammonia, methane and amino acids, water (from which most of the ice is formed), hydrogen sulphide, hydrogen cyanide and cyanogen.[26] It could, according to some reckonings, contain lumps of rock the size of a small house,[27] weighing possibly millions of tons. The closest that scientific instruments have come to examining a comet in flight was the satellite Giotto which rendezvoused with Halley on its return to Earth in 1986. Giotto found that the nucleus was more solid, darker and hotter than was expected. In fact many comets are not visible to observational astronomers because they have a coating of black dust that makes them too dark to reflect radar beams. In addition some comets are pulled into tight orbit around the Sun, and could line up in front of it, becoming virtually invisible.

Certainly Halley was massive by cometary standards. The energy of a Halley-type comet is a billion times greater than the largest nuclear bombs ever tested. A 400 million megaton hit of a Halley object would produce a 50 km crater.[28] But the chances of this happening would occur only once or twice in a billion years.[29]

Scientists are not absolutely certain of the origin of comets. They may well be remnants of the early solar system formed from the original dust cloud. There could be as many as five million comets in existence, vastly more than the known number of asteroids, mostly trapped in a huge oblique orbit some 18 billion miles beyond the sun, and known as the Oort cloud after its Dutch discoverer. The outer edge of this cloud is at least twenty times more remote than the edge closest to Earth. Oort demonstrated the comets, nearly 100 billion of them, were from a region nearly one light year away.[30] Richard

Muller raised the number of comets to 10 trillion, and Mark Bailey, of Sussex University, believes there might be another swarm of comets much closer to the Sun. The Oort cloud is occasionally replenished with new comets when the Sun moves through the spiral arm of the galaxy.[31] They are easily perturbed, and this explains why many experts think it more likely that it was a comet that was nudged towards Earth to cause the K-T extinctions. Only gravity, of course, could make giant cosmic objects follow great hyperbolic orbits, and which could at the same time display evidence of periodic cycles of such enormous duration.

In fact cosmic theories seem to be desperate and driven solutions. Great orbits and astrodynamical functions are involved in order to accommodate as rationally as possible the baffling periodicities. The path of a comet can be almost any kind of ellipse, and can change while still in orbit and become a parabola or even a hyperbola. They can strike virtually any of the planets. Nevertheless, because of the vastness of space their potential targets are not easily sought out. In the case of a comet following an elliptical path it becomes part of the solar system, but is only visible when it periodically comes close to Earth. When it passes the Sun and returns to the outer solar system it is considerably reduced in mass. On a subsequent return past the Sun yet more of it is evaporated away until it is whittled down to its central core of rock. Of the 700 known comets most of them do have periodic orbits. About six new comets are discovered every year, hinting at how many there must be in the solar system.

The reason why comets are confused with meteorites is the theory that comets are responsible for meteor 'showers', for when a comet begins to break up, minute fragments spin into their own orbits, forming a belt trailing along behind the comet. Shooting stars are said to occur when the Earth passes through this Oort ring. The rings around Jupiter and Saturn are said to be bits of broken-up comets sucked into gravitational orbits.[32]

Is There a Planet X?

The Alvarez team published their first paper on 6th June, 1980 in the prestigious American journal *Science*. It received a great deal of publicity, most of it positive, appealing to the popular imagination fed for years on catastrophic theory and extraterrestrial visitations. Some American papers said it was definitive.[33]

From then onwards the debate gathered momentum. Another radical hypothesis was advanced by Daniel P. Whitmire and John J. Matese of the University of South-Western Louisiana, who suggested that there is yet an undiscovered tenth member of the solar system. It is this planet, suggested by some to be out beyond the orbit of the outermost member, Pluto, that accounts for the anomalies in the movement of the other outer planets such as Uranus, and possibly Neptune. And it is said to be Planet X which dislodges the comets to send them hurtling to Earth.

Pluto itself might have been the tenth planet. The discovery of Neptune in 1846 was supposed to explain the perturbations of Uranus, but the later discovery of Pluto in the 1930s was also supposed to do this. However, one major reason for believing in X is the recent discovery that Pluto is not a proper planet at all. Even with its moon Charon, only discovered in 1978, it would not provide the necessary mass to cause the perturbations of Neptune and Uranus, which seem to meander more than they should if subject only to the gravitational influence of other solar planets. Recent determinations of the size and mass of Pluto, using IRAS (the Infra-Red Astronomical Satellite), show that the combined mass of Pluto and Charon puts it at 1/439th Earth masses, or about one-fifth that of our own moon. Pluto itself is only one-eighth as massive as the moon. It is almost certainly, then, a mixture of rock and ice spun off from a parent planet, probably Neptune.[34]

The most likely remaining theory is that Pluto could be discounted and replaced with a large planet, with a mass some five times that of Earth, but with its density unknown. It could have an orbital period of only about 1000 years,[35] and would probably change its orientation slowly in space. Whitmire and Matese suggest that Planet X is undetected so far because it has an unusual inclination in regard to the other planets, thus misdirecting all earlier searches. In an article in *Nature*, Whitmire and Matese said that the planet could be made to precess, and could pass through the Oort cloud twice every 52 million years on an orbital journey taking it once round the Sun every thousand years, loosing off comet 'showers' every 26 million years.[36]

Soon the critics got to work on the Planet X theory. In order to fulfil its required astrodynamic function it would have to be enormous. According to astronomer Robert Harrington, of the US Naval Observatory in Washington, it could be another Jupiter, with a mass three to five times that of the Earth. To be able to influence Uranus, he believes, it would have to incline to the plane of the solar system at an angle of about 30 degrees and have a great elliptical orbit. X's ability to function as required — to punch out comets in sharp bursts at regular intervals — was also questioned. Any perturbation from a great distance could cause them to expand the plane, triggering comets much more randomly;[37] i.e. the storm would be long and drawn-out rather than short. Its weird careening 30-degree trajectory — stretched no doubt by astronomical imagination — again made it an unlikely candidate. And its 52-million-year orbit round the Sun hardly makes it a normal planet!

Even so, the Planet X theory was soon being partially rescued by empirical observation. All the major spacecraft like *Pioneer*, *Mariner* and *Voyager* have highly gravity-sensitive instruments. They could easily detect a very large solar planet if it exists, and important clues were forthcoming from data sent back by *Voyager 2* during trips to the outer planets. John Anderson, a National Aeronautics and Space Administration (NASA) scientist, told a news conference at the Ames Research Centre in July 1987 that data from the earlier *Pioneers 10* and *11* showed irregularities in the orbits of Uranus and Neptune.[38] And late in 1989 in the aftermath of its bypass of Neptune and its giant moon Triton, it was reported that Triton orbits Neptune in the opposite direction to that of the other large moons. Its surface features were also strangely uneroded.

These facts, together with the perceived irregularities in the orbits of Uranus and Neptune, have suggested to both Robert Harrington and Mark Littmann of Loyola College, Baltimore, that Planet X could exist. Littmann suggests that perhaps Pluto was once a moon of Neptune's which was broken in two by an 'intruder' (with Charon becoming the smaller fragment) which then expelled it to the edge of the solar system. This would explain Pluto's highly eccentric orbit.[39]

The Death Star

If Planet X has astrodynamic weaknesses the Death Star theory had even more, since it was seriously suggested that what disturbed the comets or meteorites at regular intervals was an invisible dim companion star to our own Sun. Oort realized that most orbits in the inner solar system are kicked there by passing stars.[40]

The Death Star theory is based upon the fact that most stars in the galaxies are found in pairs, while the Sun is alone. More than half of the stars we can observe with telescopes are binary or multiple stars, i.e. they have pairs or groups of physically related suns. The formation of stars from primordial nebula seems, in any event, to lead to star multiples as a matter of course. Seen through even a low-power telescope a very high proportion of stars are double or multiple systems, orbiting around a common centre of gravity. There are even clusters of four stars which can be referred to as 'multiple binaries', with the pairs being widely separated. The bright star Castor in the constellation of Gemini (The Twins) is in reality a splendid interrelated system of six separate stars.

In a triple system two of the stars can lie quite close to

each other, with the third member much further out. With Alpha Centauri, for example, our nearest stellar neighbour, Centauri A and B lie twenty-four astronomical units (AUs) apart (over two billion miles), and have a revolution period of eighty years. According to some astronomers a third member is a 'cool' Class M red dwarf star (a dwarf star is one with its nuclear energy spent), orbiting the other two at a distance of a tenth of a light year. It comes closer to us than Centauri A and B, and is therefore known as Proxima Centauri. But, like Proxima Centauri, the further out the multiple or binary star is, the less likely, given the limitations of astronomical telescopes, are scientists to agree that it is part of a cluster.

Some of the component stars can also be of distinctly different colours. A particularly striking example is the red supergiant Antares (Alpha Scorpii) in the constellation of Scorpio, near the end of the southern horizon. Antares can be seen with a lovely green companion star, and the two have been described as a 'celestial ruby and a celestial emerald, set in nocturnal velvet'.[41]

Earth's hypothetical twin sun was dubbed the 'Death Star' by its supporters, and is occasionally known as Nemesis (after the Greek goddess of retribution). It is thought to be about 2.5 light years distant from the Sun, and could well be a tiny red or brown dwarf star. It is worth remembering, however, that some seven out of ten of the nearest stars are red dwarfs.

The Death Star is the brainchild of Richard Muller and Marc Davis of Berkeley. Muller had always been intrigued by the dinosaur mystery,[42] and very nearly, at an early stage in his career, became a paleontologist instead of a physicist. Muller was aided and abetted by Piet Hut, a Dutch physicist at Princeton University, who suggested the companion star passed through the comet belt rather than the asteroid belt. Hut said that even if the companion star came within half a light year of the Sun it could perturb most of the comets. It would more likely be a comet storm, with Nemesis knocking out many of them in a vast elliptical orbit, lasting anywhere from one to two million years. In their Nemesis paper, Muller *et al* said the orbit should vary by about 10 per cent because of perturbation of passing stars, and this would go along with episodes of cratering on Earth.

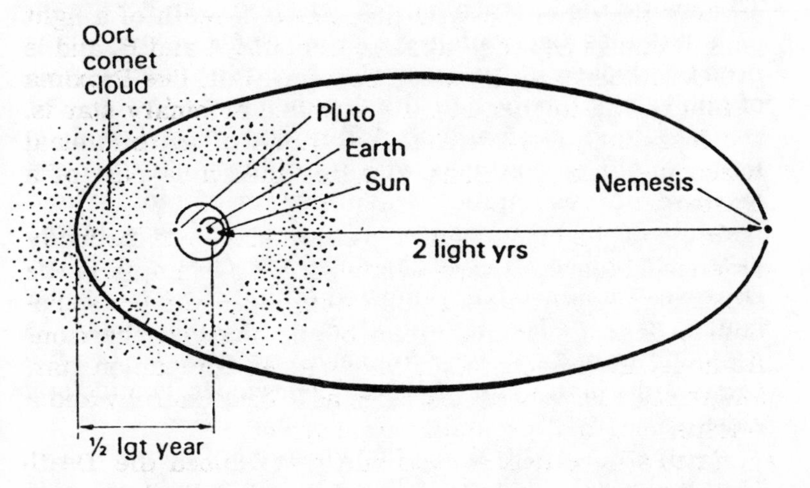

Fig.22. The Nemesis Theory. Taking some 26 million years to complete one orbit, the faint star Nemesis passes through the Oort cloud to dislodge comets, sending some of them hurtling towards the solar system. (Source: Quest *magazine, October 1984)*

One problem concerns mass and luminosity. The rate of production of kinetic energy varies with the cube of a star's mass. So Rigel,* only twenty times 'bigger' than the Sun, yields 10,000 times more energy per second, and

* Rigel lies in the Orion constellation, and is 900 light years away.

this affects its luminosity. This poses problems for the Death Star, since its luminosity must be extremely low and its distance from the Sun great, about 2.6 light years away. So, based on our comparison with Barnard's star,* Nemesis must have a mass no greater than 1.2 times that of Barnard, or about 0.12 solar masses.[4 3] It would hence be about 10,000 *million* times fainter than the full moon. This makes the improbability of detection even greater.

It has also been suggested that Nemesis may have been overlooked in the mass of stars on the outer fringes of our galaxy, about half the distance away to the next closest, Proxima Centauri. A dim star, says David M. Raup, might be confused with a bright star further out because no systematic measurements of motion and parallax have been made (because few believe in the Death Star theory), to detect a small star close to another. Because of its unstable, elongated orbit a close encounter could easily deflect it, and thus once again make its detection from Earth very unlikely. Nevertheless, bearing in mind the limitations of existing telescopes, it could still exist.

Yet there is still the well-known stability problem. The Death Star would need to have an unusually massive orbit, with the Sun and the twin star being three light years (18 trillion miles) apart. If it was formed at the beginning of the solar system, its orbit would become ever wider. In principle this means that the twin could easily have been knocked out of its orbit aeons ago by a passing star, or even a dense part of the dust cloud, a theory advanced by Richard B. Stothars and Michael R. Rampino of the Goddard Space Institute. Or it could have been driven into a much tighter orbit. Stars in a double-star system are known to be more than a third of a light year apart. In theory other twins could be much further apart, but risk greatly increasing the instability factor. In

* Barnard's star (named after its discoverer) is the second closest star to the Sun, and is roughly six light years away.

fact it would have less than one chance in a thousand of surviving its orbit; an orbit which would, in any event, have to be 'eccentric' or greatly stretched.

A further difficulty with any possible twin sun is the position of Jupiter, very like a failed sun itself. If the twin star had the same mass as the Sun, it would have to take the place of Jupiter (5.2 AUs from the Sun, according to computerized studies). It would hence dramatically disturb the orbits of the other solar planets, with Mars being sent spinning into space forever. Earth's orbit, however, might remain stable even if our pair of suns were only 0.2AUs (about 18.5 million miles) apart.

Some astronomers remain convinced of Nemesis's existence because of the unmistakable evidence of sources of galactic gravitational forces which periodically seem to shift their position. Astrophysicist Armand Delsemme of the University of Toledo in Ohio has discovered that the paths of 126 comets orbit the Sun in oddly skewed configurations. And then there are the perturbation of the outer planets and the knowledge that some twin star systems very close to Earth (like Proxima Centauri) were only observed more or less by accident. There is still a continuing search for a reddish dwarf star not more than three light years away, despite the disappointment caused by the failure to spot anything unusual in 1983, after sifting through the recorded transmission of the IRAS infrared satellite that revealed valuable information of more than 250,000 cosmic objects.

The ET v Terrestrial Debate — The Implications

And so the debate continues. Both sides claim that the fairly widespread global distribution of the metal grains could have come from volcanoes or a bolide, since in both events they would be blasted upwards from the surface with tremendous force as part of other gases and debris. Both the geochemical evidence and the physics involved in spreading the evidence are still the subject of controversy.

The terrestrial evidence is not foolproof. There are wide variations in literally millions of years in the timing of the extinctions with known periodicities, and with crater impacts. There are also many errors and approximations in regard to the mooted orbits and trajectories of these new hypothesized cosmic bodies which add up to millions of miles. There are many other uncertainties, for although volcanoes could be responsible for the unusual osmium isotopes, which are related to iridium, the actual quantity discovered suggests it was more likely to be deposited by a 'large impact'.

One theoretical difficulty arises from the fact that iridium is more easily detected than the element in the same chemical group, because it can be isolated by neutron activation at much lower concentrations. On the other hand, the noted variations in the density of discovered iridium from sites around the world could be proof of volcanism, although the blatant disagreements about the amount of iridium detected from modern volcanic eruptions (as witnessed by the contrast between Frank Asaro and William Zoller's findings) suggests much more research needs to be done.

The problem of how the iridium came to be concentrated along or close to the K-T boundary has also been raised. Various hypotheses focus on the concentration of particles from bottom currents surrounding rock strata that become chemically eroded and later enriched with iridium, or to geochemical processes in an early terrestrial environment with an oxygen level different from today's[44] (*see* Chapter 9). If the erratic behaviour of volcanoes had been responsible, then it must be borne in mind that particulate matter usually remains suspended in the atmosphere of the hemisphere into which the volcano discharges itself. The Deccan Traps were in the southern hemisphere some 66 million years ago, and some scientists say this is why the concentration in the south (like that of New Zealand) is higher.

Other scientists have said the fossil record has been misread. Jan Smit and S. Van Der Kaars of the University of Amsterdam contend that the floodplain had washed away dinosaur remains, so that specimens from the Paleocene were mixed with the Cretaceous, thus seeming to make modern mammals more numerous.[45] In reply the gradualists say that the Amsterdam researchers were focusing on an area too far away from the main K-T excavation site.

Adherents of the ET hypothesis, in the meantime, weaken their case through equivocation. Michael Rampino muddies the waters by admitting, in a *New Scientist* article, that some features of the boundary do, in fact, have a distinct volcanic aspect.[46] He says he found weathered clay minerals 'similar to those typically found in beds of altered volcanic ash'. Certainly the presence of osmium could be construed either way, he maintains. He even blurs the distinction between the volcanic and ET argument by saying an impact could have triggered massive eruptions of 'flood basalt' (lava) by stripping away the parts of the crust and upper mantle to expose hot rock below. Hence weak points could occur where volcanoes could force their eruptive way to the surface. It is fair to point out that other scientists, in the absence of proof, do not accept this explanation.

Rampino then goes on to talk about the chance of a massive Earth crater between 100 and 200 metres wide, as would be expected from the sort of large celestial object needed — as proposed by the Alvarezes. Possibly, he says, a large comet hit the ocean and the crater has since disappeared down an ocean trench, perhaps somewhere near the original Deccan deposits. Alan Hildebrand, of the Lunar and Planetary Laboratory of the University of Arizona at Tucson, suggested the missile crashed into the sea east of present-day Nicaragua. This is based on what appear to be deposits left behind by a violent seismic wave — a tsunami.[47] Certainly, a 100-200 km crater would gouge out crust up to 40 km deep — and the resulting lava outpourings, if the missile struck land, would be more than enough to obliterate the crater.

Further analysis done by Michael Rampino with Richard Stothars of Goddard, suggest that eleven important eruptions of flood basalt over the past 250 million years closely match the times of mass extinctions *and* the cluster of impact craters.[48] His argument is that an impact can expose a subterranean hotspot which can give rise to vast outpourings of hot rock, and most of these have occurred historically as a result of continental drift. But occasionally such hotspots may have been caused by violent impacts, with the same flood-basalt consequences.

Even the impact might not be necessary, according to another theory emanating recently from Florida State University. David Soper and Kevin McCartney say that a plume of molten lava rises through the Earth's 100 km mantle to form a hotspot volcano, leaving behind a thinner layer which gradually thickens until it becomes unstable. Soon, massive explosions occur when the hot magma hits the cool base of a continent. Then the rapid cooling of the molten mixture could produce enormous pressures that could form the patterns of multiple shocks in quartz crystals.[49]

There are still problems, as research continues, with the question of magnetic fingerprinting and with establishing a consensus about the length of geological time appropriate for the boundary clays, something that will plague us when we later try to place the death of the dinosaurs within a specific time zone. Did the disaster occur abruptly, and did it also occur around the world simultaneously?

The complexity of the argument, and the continuing lack of consensus, was revealed at a meeting on global catastrophes in Earth history, held at Snowbird, Utah in October 1988.[50] The absence of a conclusive theory, far from leading to fractious argument, has stimulated the scientific community who appear to thrive on controversy. Several major puzzles remain: one is whether the K-T extinction is unique. Many scientists believe not. For example, David Bice of Carleton College in Minnesota has recently found evidence of shocked quartz at the boundary between the Triassic and the Jurassic in Tuscany, Italy. Only an impact, it is reckoned, could have caused the three separate sets of flaws in the tiny crystals. It is curious to note, however, that one other puzzle for impact extinction theorists rests on the lack of iridium in such craters as the 70 km wide Manicougan crater in Quebec, reckoned to be some 211 million years old. This crater is often widely cited as true evidence for a late Triassic impact.[51] The K-T missile, incidentally, has two favourite impact sites: one is beneath the coast off Yucatan in Central America, as evidenced by what looks like deposits from a tidal wave in Texas;[52] the other appears to be beneath the fields of Iowa, where a circular pattern of rock is buried below many metres of Ice Age sediment.[53]

It is worth mentioning here that Colin Patterson of the Natural History Museum in London says that Raup and Sepkoski's conclusions are an illusion based on imperfect data. At a meeting of paleontologists in Boulogne in October 1988, Patterson claimed he had gone over Sepkoski's compendium and found many of the fish and echinoderm extinction records were wrong, because the animals themselves had been mis-classified by the hundreds of researchers who had compiled the list, or were influenced by habits of recent fossil collectors.[54]

In the meantime, Steven Jay Gould of Harvard summed up the Snowbird meeting: 'Impacts add an enormously quirky character to life's evolution.' In evolutionary terms this means that if creatures are to survive the kind of fate imposed on them by scientists, they would surely need to have different characteristics.

Kenneth Hsu points to the irony of the catastrophists being challenged to prove that the extinctions had been rapid: 'Now that a global catastrophe is largely accepted, it turns out that extinction was probably not so rapid after all. We are now challenged to explain how it could have been so slow.'[55]

Finally, the debate has, in all likelihood, shifted intellectual parameters concerning the ancient history of the Earth. The term 'neocatastrophism' is now being used to suggest a revival of eighteenth-century catastrophism which dominated geology. Charles Lyell's substantive uniformity postulating that events have always unfolded themselves at the same 'intensity' as the present is now recognised as merely an a priori assumption that can now be modified in the light of more recent discoveries and interpretations. Perhaps, as T.F. Malone suggests, following the revolution in the earth sciences brought about by continental drift some twenty-five years ago, we are now in the midst of a second twentieth-century revolution in the planetary sciences.[5 6]

The controversy has reached many diverse academic disciplines, such as geology, paleontology and evolutionary biology, even posing a threat to Darwinian theory. The ET theory could never be proved beyond all doubt, but being testable it aids understanding even if only as a working hypothesis. Death will not be due to conventional evolutionary means involving maladaptation, food supply, species competition, etc. Instead it will seem disturbingly random.

Chapter 8
WHY EARTH DESTROYS LIFE

THE ET/TERRESTRIAL argument continually risks posing a false explanatory dichotomy. The giant bolide that is capable of devastating Earth would be powered literally by energy from the cosmos. And the kinetic forces it would unleash and the atmospheric turbulence it would create on impact would not, therefore, have been 'home grown'. The most obvious manner in which the two approaches are unified is when the ET event triggers other violent Earth events: for example, when an asteroid sets off a chain reaction of volcanic eruptions.

In 1990 a multi-causal explanation for the K-T extinctions was advanced by Princeton paleontologist Gerta Keller, who studied the remains of single-celled marine organisms at sites in Tunisia, Spain, Denmark and Texas. She found, incidentally, that many species died out some 300,000 years before the K-T, and many others continued to exist a long time after it, although in reduced numbers. She thinks scientists searching for the cause of the K-T extinctions should not neglect the effects of lowering sea levels, global cooling, and increased volcanism, possibly in conjunction with the impact theory.[1]

One other typical bridging theory looks at first glance suspiciously like a pseudo-science theory. The 'pole shift' theory focuses upon a mobile and fluctuating Antarctica, which is viewed as a giant gargoyle at one end of a smooth sphere that is tilted away from the upright and hence could literally flip the world over, with the south pole becoming the north.

Emmanuel Velikovsky, a Russian expatriate psychologist who wrote several best-sellers on this theme in the 1950s and 1960s, was a rather more erudite pseudoscientist than most. He took brilliant advantage of both the historian's lack of secular knowledge of supposedly true 'Biblical' events, and the early geographer's rather speculative understanding of early Earth history prior to the revolution brought about by plate tectonics. He then painted an alarming catastrophic picture of events occurring several millennia before the birth of Christ. The doomsday message of the Old Testament, many parts of it couched in apocalyptic and tumultuous language, with floods, earthly violence, plagues, and Godly wrath being meted out to sundry tribes and races, was taken at literal face value by Velikovsky. He pushed the explanatory logic of the early impact extinction theories to their utter limits, to the embarrassment of both clerics and mainstream scientists.

Even so, Velikovsky's comet was grotesquely massive even by science fiction standards. Ripped by tidal force out of the body of Jupiter, it was a red-hot bolide almost the size of Mars! After roaring past the Earth, causing the planet to tilt and bringing about weird atmospheric disturbances, the north polar ice-caps were shifted further north. Enormous tidal forces hurled entire oceans over land masses, while massive seismic activity made whole mountain ranges collapse. The giant comet careened on in its orbit to become the new planet Venus, which today is hot because it is, after little more than 10,000 years, still cooling down.[2]

This is not the place to list all the scientific errors in Velikovsky's works, as this has been done admirably by the astrophysicist Carl Sagan in his book *Broca's Brain.*[3] We mention Velikovsky here because he was the pastmaster at expounding the discredited but entertaining 'pole shift' theory, where the world could have been spun over on its axis. This is a surprisingly persistent theme among occultist writers, largely, one suspects, because it can be presented as a modern-day validation of 'cometomania' — the eighteenth-century fear of comets as harbingers of Earthly doom and destruction. It also, unlike many theories on the occult, appears to invoke perfectly plausible (at least to those unversed in astrophysics) physical laws, rather than mystical or supernatural forces.

It is, incidentally, never a good idea for mainstream scientists to simply dismiss pseudo-scientific arguments without debate. Earth science is going through a revolution in understanding right now, with imaginative theory only tenuously related to the empirical evidence needed to support it. Many of the ET ideas discussed in the previous chapter have more than once crossed over into the realm of erudite speculation. For another, it is better ethics to benignly confront unorthodoxy with what is proven and known, conceding only what is possible within physics, and to point out why no further progress can be made with such theories. Humility is always needed, because there is still a likelihood of a brilliantly unorthodox hypothesis being validated in quite unexpected ways. 'Continental drift' theory, as we shall see, was once considered to be a pseudo-science.

Take the pole shift explanation again. Researchers at the Californian Institute of Technology (Caltech) in 1987 had modified and revived an earlier theory first advanced by the eminent physicist Thomas Gold. Gold suggested in a *Nature* article in 1955 that the Earth had rolled over in slow motion several times in the past. If a continent the size of South America, he surmised, were raised 30 metres by tectonic movements (of which more later), the axial spin and the axis of angular momentum would cause the planet to topple over at a rate of one degree per thousand years.[4]

Similarly, the Caltech researchers suggested that the early break-up of land masses starting 400 million years ago may have destabilized the Earth's rotation. And it is known that the uneven distribution of Earth's land surfaces can indeed create frictional resistance which over centuries has slowed down the rate of Earth's rotation. The day has actually grown longer as a result — in the last 400 million years the day has lengthened by 6,400 seconds, or almost 1.8 hours.

The Importance of 'Tidal' Forces

But tidal (in other words, cosmic) forces also play a part in shaping the physical world and creating turbulence in the world. And it is worth bearing in mind that proponents of tidal/gravity forces themselves were once criticised as 'fringe' scientists. The universe is dynamic, with the Sun, Earth and the galaxy itself performing constant orbits, and propelled by kinetic, EM and gravitational forces. Much of the variations in heat and cold in celestial objects, including the solar planets, are the products of trillions of such energy cycles. In a sense these cycles are little more than information codes emanating from EM rays. We now have new theories about why the moons orbiting the outer planets like Jupiter and Uranus have cracks and ridges: it is because of tidal and EM phenomena that make heat build up inside. Moon orbits are often pushed and pulled by the mother planet, and the moon-satellite absorbs energy within a sophisticated resonating relationship from its sister moons. Planetary physicists are overhauling their bodies of knowledge; since volcanic eruptions on solar planets and moons were thought to be entirely due to an object's mass generating radioactive decay at the core.

The solar planets, then, exert gravitational pulls on each other, or often cancel each other out. But occasionally they are aligned in a synod where their pulling power is greatly enhanced. The mass of an object, of course, greatly determines its pulling power. For instance, the tidal effect of the Earth on the Sun is only 10,000th that of both the Sun and the moon on Earth. Gravity, although still much, much weaker when measured in absolute terms against the forces that bind together the constituent parts of the atom, is nevertheless still powerful on the cosmic scale. Indeed it can be very useful because of this. The *Voyager 2* spacecraft was helped on its exploratory journey through the solar system by the aligned pulls of both Saturn and Jupiter, reducing a potential 30-year journey to just eight and a half.

An important theory of the 1970s was the Jupiter Effect, which decreed that the outer massive planets had a gravitational and EM influence on the Sun, and were partly responsible for the formation of sunspots. This theory, once thought to be dubious, has gained some support from recent researchers. For example, physicists have shown that the Sun is being tugged across the solar system's centre of gravity by the outer massive planets[5] producing bulges on the Sun's outer photosphere as it performs its own small elliptical orbit at the centre of the vortex. Physicists at Denver University have also shown that showers of high-speed protons are caused by certain planetary line-ups. K.D. Wood of Colorado University calculated that the pull of Jupiter, Venus and Earth probably caused the giant flare that was observed across the world in 1972. Others have pointed out that the 11-year sunspot cycle is the same as Jupiter's orbital period.

For this reason many earth scientists have suggested that detected periodicities on Earth, such as Ice Ages, or the extinction of species, are somehow connected with these cosmic periodicities either via the Sun or through other cosmic features, such as meteoritic bombardment. The Sun in turn is dependent upon what goes on in its surroundings. And all biological life is utterly dependent for good or ill upon the Sun's radiation. That Ice Ages, or mini Ice Ages or other extreme weather phenomena, seem somehow to be connected with fewer sunspots than usual has been discussed frequently in the scientific literature during the last two decades. The drought in the Sahel, which ended with catastrophic floods in 1988, seemed to coincide with a sunspot cycle at its maxima. In 1946 John Henry Nelson, an engineer for RCA Communications, was one of the first to concentrate on the influence of sunspots on radio reception. He soon discovered that the planets had something to do with sunspots, and EM was closely involved. He concluded that the Sun was little more than a giant armature hanging in space, with the planets like celestial magnets.

The planets, clearly, have this curiously symbiotic relationship with the Sun; the fluctuations, rhythms, periodicities, having a direct feedback on the Earth itself. Many of the episodic warming and cooling periods may be no more than coincidences, like the earthquake in Morocco in 1960 that took place one year after a massive solar flare. Nevertheless it is puzzling to observe that rainfall patterns do seem to be concentrated in certain phases of the month and not in others. And avalanches and severe winters do seem also to have a 12-year cycle as a common denominator.

Let us take an example which has recently come to light in the scientific press: earthquakes in the eastern United States seem to occur in time with the tides of the Earth. This in turn has something to do with the changing shape of the planet in response to the moon's tug. Robert Weems of the US Geological Survey suggests that earthtides trigger quakes along a network of fractures formed more than 60 million years ago. Most of the recorded quakes happened after a 'high tide' in the body of the Earth,[6] and were exacerbated by a Sun at a declination between 17°N and 17°S.

Violent episodes of volcanism may instead have been due to the violent contraction as the Earth slowly crumples and shrinks. This is the view of Raymond Littleton, of the Institute of Astronomy, Cambridge, who said that the Earth is decreasing in its radius at a rate of one-tenth of a millimetre a year, shrinking by a total of about 300 km, as a result of the remorseless rise in internal radioactive heat which literally eats away at the solid internal parts of the planet. At 100 million-year intervals, he suggests, periods of shrinkage and resulting volcanism probably occurred.

When the Earth is contorted in this way, i.e. either by radioactive or cosmic events, other knock-on events occur. When quakes redistribute the internal mass of the planet this could affect the length of the Earth day (LOD) by making the Earth more compact, and hence in turn making it spin faster. Studies from the Goddard Space Institute in 1988 suggested that of 2000 quakes they had monitored, the change in the LOD could have been altered by at least one percent. Storms and atmospheric disturbances have also been said to have triggered a slow-down in the LOD. To complicate matters, some scientists say there is a solar connection with LOD, with volcanic activity reaching a peak when pronounced changes in earthspin were observed.[7] This is because Earth's rate of rotation has not always been smooth and even; often the Earth has juddered with internal seismic activity. Earth has had an extremely violent history, especially millions of years ago. This has not only affected the rate of spin, but the spin itself when speeded up or slowed down has also affected seismicity in a feed-back process. And an ash-laden atmosphere may also in turn have slowed down the Earth's rotation in yet another feedback loop.

The most obvious and well-understood Earth cycle concerns the ebb and flow of the tides. The moon moves in its orbit in the same direction the Earth spins. The oceans become heaped up under the pull of the moon, giving diurnal tides. When the much smaller influence of the Sun (smaller because it is 390 times further away than the moon) is added in, other additional mini-tides are created, often known as 'spring' tides. These tides have caused devastating floods when swollen seas rush through a narrowing estuary to become a tidal bore.

The Sun is, of course, the prime instrument in goading the entire 'weather machine' on Earth into action. Much attention, as we have seen, has focused upon the peculiar role of sunspots in bringing about episodes of hitherto inexplicable or prolonged and extreme weather. Now the scientific trend seems to be moving away from analyzing sunspots in particular to solar 'activity' in general. (The Sun's overall diameter does vary marginally, and its radiation spectra — including gamma rays, X-rays, UV and infrared rays — can also vary; but we must be careful not to say that the Sun merely gets 'hotter'.)

In the meantime solar physicists guess that many of the effects of the Sun work through EM rather than infrared heat. Sunspot cycles can affect the west equatorial winds and atmospheric pressure in the northern hemisphere. Goesta Wollin, of the Columbia University, New York, has predicted snowstorms in America following sudden changes in the strength of the solar field, in which Earth's field is embedded.[8] But the difficulty in rigorously tying up EM fields with changing weather patterns is the way in which Earth's field often varies both historically and geographically, and both weakening and strengthening of the field can have either a climatic cooling or warming depending on a complex interrelationship of magnetic-meteorological factors. One theory, advanced by the distinguished astrophysicist William McCrea, has the Sun crossing the spiral arm of the Milky Way galaxy, where cosmic matter is more congested. The Sun begins to absorb dust into itself, and in a sense 'warms up'. But objects much closer to the Sun can affect its output, and the amount of radiation from the Sun (and not just its IR rays) can be varied considerably in this way.

The Nature of Terrestrial Energy

It is usual to describe a number of types of energy that differ from each other in the way in which they constitute the physical reality of our universe. It is significant that scientists, understanding the laws of physics as they apply to human societies, often rank these energy forms from the terrestrial to the cosmic: first we have mechanical energy, then thermal, electric, chemical and finally energy-in-matter.

It is from energy-in-matter, in a real sense, that all other energies are derived, and this is the reason why the opening chapter of this book talked about the formation of the solar system, and Chapter 7 dealt largely with cosmic and solar-system energy: according to Einstein's famous formula, all matter consists of concentrated energy: $E = Mc_2$. This equation has enormous importance for all branches of physics, and, for that matter, all branches of science. $E = Mc_2$ tells us that mass at rest has energy, but when a particle of matter becomes all energy, say when it becomes violently heated in some cosmic situation, or when it encounters an anti-particle (every particle is said to have an equal but opposite anti-particle, with the same mass but with a reversed spin and electric charge), it has rest mass. It then becomes radiation that moves at the speed of light.

For our purposes it is enough to say that energy is created when differently charged atomic particles collide with each other. Working backwards, then, we can say that chemical energy is released after being stored in the linkage of atoms to molecules. Compounds can vary in their energy richness. A typical energy-rich substance is a fossil fuel, either solid or in liquid form, which is released in a chemical reaction with oxygen, when part of the chemical bond is converted into thermal energy and carbon dioxide as atomic particles collide and interact with each other. Both electrical and mechanical energy can also create thermal energy. It is interesting to note that Man can make things go against the grain of Nature by manipulating energy flows to his own advantage. Take refrigeration. In this we seek to transfer energy, as heat, via an engine of sorts, from hot to cold at the expense of electricity generation. This generation has the effect, of course, of increasing the amount of energy used that in the end only increases both local heat and universal entropy. It is this post-Einsteinian understanding of matter-in-energy, in other words, that is important to our discussion of the relationship of dinosaurs with the biosphere, a much more complicated energy machine.

Let us return for the moment to chemical energy. Chemical reactions, in the atmosphere and elsewhere, are heavily disguised heat engines. Chemical reactions can produce heat, but not in the same way nuclear reactions can. The controversy about 'cold fusion' that arose early in 1989 widened an age-old gulf between the chemist and the physicist, each jealously defending his territory. Traditionally chemistry stops at the perimeter of the atom's nucleus, where nuclear physics begins. What Martin Fleischman of Southampton University and Stanley Pons of the University of Utah had concluded from their test-tube studies, in a blaze of worldwide publicity, was that energy had been created, similar in principle to that arising from nuclear fusion experiments, from a purely *chemical* reaction. Fusion occurs when two atomic nuclei meet to form another, heavier, nucleus. Massive amounts of energy are released, and tiny nuclear particles, neutrons, burst out; a sure sign that fusion has taken place.

In chemical reactions one substance changes to other substances, but the nuclei remain the same. The mass-to-energy ratio, compared with nuclear reactions, is minute: pound for pound, nuclear fusion gives out about a million times more energy. Pons and Fleischman claimed their cold fusion even took place at room temperature — a far cry from the searing heat of 100 million °C required in the classical notion of fusion. Yet both claimed to produce the excess heat which is the hallmark of fusion. Critics, however, pointed out that the heat generated was either of the same order as the electrical energy going into the experiment, or background radiation was not taken into account, or their experiments were not rigorous enough; i.e. they failed to control experiments with 'light' water.

Nevertheless, what chemical reactions *can* do is rearrange chemical molecules, using the same principle of a descent into chaos, simply by dispersing energy. This is why it is so easy to misunderstand what is happening in chemical experiments. The physicist Michael Shallis points to a block of iron surrounded by oxygen in the atmosphere.[9] As rusting takes place the oxygen in effect cools the metal by an infinitesimal amount when it combines with the iron, and energy is released as heat.

Food, as a product of a stream of never-ending corruption, by nevertheless providing energy for animals is similarly a kind of chemical heat engine.

Heat, however, is like a bad currency. It can seldom be exchanged for any other kind of energy, unlike all other forms. We can say in a nutshell that energy can be turned into work only if, within the particular system we are using, there is not an even flow of the concentration of energy. In other words heat, as a kind of energy in a less concentrated form, arises after conversion from energy originating at a higher point of concentration.

The Second Law of Thermodynamics is usually attributed to the French physicist Nicholas Carnot, who, in 1824, made important advances in the understanding of heat. He was probably the first to appreciate and explain the theory behind the steam engine. In the late eighteenth century, prior to the understanding of thermodynamics, heat was thought to be the product of vibration. The American traveller and physicist, Benjamin Thompson, supervised the boring of cannon in Bavaria in 1798, and noted that quantities of heat were being produced. An even earlier theory held that heat was a material substance that could be poured from one vessel to another. It was named caloric, from the Latin for 'heat'.[10]

It was only with the understanding of the atomic nature of matter that the notion was born, of heat arising from molecules in continual motion in a gas, bouncing off one another.[11] This was later formulated as the *kinetic theory of gases* by James Maxwell and Ludwig Boltzmann in the mid nineteenth century. It was noted that the rapid vibration of molecules in substances could be speeded up by the addition of yet more heat. Even solid metal could become molten when the bonds between neighbouring molecules were vibrated sufficiently violently to break them, so that molecules could move freely past each other. Heat, then, was the total energy contained in molecules of a given quantity of matter. Every chemical compound had energy stored within it in the atomic bonding forces. It was when these bonds were broken that the energy escaped as heat or light. The burning of coal involves breaking the bonds between the carbon atoms in the coal and those in the oxygen molecules with which it combines to produce carbon dioxide (which would then have less energy in its bonds). Thus grew the new science of chemical thermodynamics.

Earth as a Heat Engine

Let us now broaden our theme to look at Earth itself as a heat engine. It is interesting to note the curious polar symmetry of the creation of heat balances in solid planets without atmospheres like our own. First the planet is formed, as we have seen in the first chapter, by cold accretion. Soon, however, the cold welding becomes hot welding. Although the initial energy of the infalling particles is low enough to keep temperatures at under 100°C at the surface, and to allow water vapour to develop, denser materials much further down create a great deal of heat. Planets and moons with a radius greater than 1000 miles across, bearing in mind what we have said about tidal forces, necessarily have molten hot cores, with the lighter silicates, sulphur and water vapour being outgassed through volcanic eruptions. Even Uranus's moon Titania is known to have volcanic activity. However, smaller spheres can lose their internal heat by radiating it away from the surface. It is clear that planets like Earth could never have been molten through and through, as this would have driven off virtually all of the volatile elements like zinc and arsenic.

The scientists at Caltech, like Don Anderson and Brad Hager, confirm what has long been known about the driving force that moves the continents around — internal heat that is not radiated away into space. The story began in the Silurian period, some 400 million years ago. At that time the supercontinent called Gondwana — made up of the forerunners of South America, Africa, India, Australia and Antarctica — straddled the South

Pole. According to Anderson a massive warm area was trapped beneath the crust (largely focusing on the spot where Africa is now), forever trying to press upwards, ultimately bringing about movements on the crust that could well be described as a kind of 'pole shift'.

Over the following 100 million years the mantle and the crust had to stretch from the South Pole to fit the equatorial bulge, thirteen miles greater than at the poles. This ultimately brought about massive rifts; one of them becoming the Mid-Atlantic Ridge. Later, about 255 million years ago, another smaller continental crust was Pangaea, and in line with the theory that continental land masses get very warm, the temperature may have reached 45°C in the centre; and the highest mean monthly temperature, 38°C, would have been lethal to living creatures. At a latitude of 30°S, summer temperatures would also have been 45°C. This temperature of the atmosphere (although not its density or composition) has been calculated by a climate computer primed by Tom Crowley and his colleagues at the Applied Research Centre. As by then plants and animals had started to populate Earth, this implied to Crowley that large parts of the globe would have been uninhabitable because of severe summer heat and winter cold.[1 2] Simulating the modern climate shows that only present-day northern Canada and Siberia would reach such extremes. Such regions cover only 8 per cent as much area as Pangaea. This may explain why so few fossils are found for the period 225 million years ago.

The main stress forces upon the Earth arise from its own violent internal dynamics. More importantly, they are responsible for its *thermal* structure. Pressure and temperature — i.e. physical rather than chemical features — explain most of the variations of the crust and below that the mantle. The crust can be up to 40 km thick in places; it is much thicker under mountain ranges. The outermost crust, together with that upper part of the mantle sliding on a weaker zone of malleability, is known as the lithosphere. The mantle, about 2,200 miles of compressed rock and hard, stable crystals, has upper and lower parts of varying densities that transmit seismic waves differently.

Recent new discoveries have been made by Harman Craig of the Scripps Oceanic Institute, which have led him to believe that the mantle is layered in a number of sub-regions, each of them having bumpy, asymmetrical boundaries but each interacting with the core below.[13] These sub-regions of concentric layers differ in density and malleability. An outer molten ball is thought to be 1000 miles in diameter. Further down, this ball is perhaps 800 miles across, consisting of superheated metal and rock squeezed by unimaginable pressures. It is the hot mantle blobs which were originally responsible for the later stresses and contractions, but these blobs would tend to reach the mantle in much less time than the great age of Earth (four and a half billion years) would suggest.

Scientists are only now gaining a rough idea of just how hot is the Inner Earth. Geophysicist Raymond Jeanloz of the University of California, Berkeley, has subjected core-like iron oxides to massive pressure, combined with laser-beam heating. He has achieved temperatures of about 6300°C, equivalent to that which most scientists speculate occurs at the boundary between the inner and outer cores of the planet.[14] Thomas Ahrens of Caltech suggests the heat at the centre is as much as 6600°C, about a thousand times hotter than the surface of the Sun, and responsible for massive explosive convulsions at the crust.[15]

Other information comes from computerized axial tomography (CAT). Earthquakes send a great number of vibrations racing through the Earth, and they tend to slow down or speed up depending on temperature and the density of the interior regions they pass through. The internal boundaries seem far less rigid than the upper ones. There seems to be a hot metallic 'seabed' upon which analogous 'continents' are floating, reflecting the drift that goes on at the surface.

The idea of a hot Earth has led some scientists to say

that although it is probably 4.5 billion years old, the first billion years of rocks seem to have been melted away without trace by the simple cumulative release of alpha particles.[16] Oceanic beds are made of basalt, which is heavier than granitic rock. It was the lighter materials that were left behind to form the outer mantle and crust, such as silicates and carbons. Slow convection of heat through radioactivity allowed the elements over a period of some 3.9 billion years, to migrate upwards to their various catchment areas. Molten specks of iron would then have sunk towards the centre, and other lighter elements would have settled further up, finally terminating with a surface in which perhaps only 6 per cent was iron-ore, with the remaining 30 per cent of the Earth's stock of iron residing at the core. Some of the heavier elements, such as uranium and thorium, could combine with other elements such as oxygen and silicon, so they too could rise to the surface. The surface was covered with many more hot-spots and volcanoes than we know today, issuing billions of tons of lava and gases.

Our Mobile Earth

How these lavas and gases reached the surface can be explained by the science of plate tectonics, as can large-scale crustal movements, the uplifting of mountain chains and the internal architecture of the planet. Plate tectonics proceeds from continental drift theory, a long-suspected physical process based upon the conclusions derived from an examination of the outlines of the continents of a model globe. It was first popularized by the German meteorologist Alfred Wegener in 1915, but it languished for years in the realms of pseudo-science. Apart from the lack of any recognised mechanism for bringing drift about, it was known (through astronomical reasoning) that the Earth's diameter was not increasing, which expanding continental masses would imply. The Caltech researchers pointed out, however, that the uneven dimensions of the Earth's mass can stretch drifting continents literally to breaking point.

Opinion shifted towards continental drift after the Second World War, gaining rapid momentum in the early Sixties. It was finally accepted by the scientific establishment when the first generation of modern computers could describe accurately how the coastal margins of the continents of the world could fit together. In fact the creative use of physics and mathematics greatly improved our understanding of the Earth. The new science of plate tectonics virtually killed off 'geology'. It was as much a cultural revolution of the 1960s as a scientific one, and completely overhauled the perspectives, understandings and methodology of all those concerned with the 'earth sciences'.

The idea that rigid land masses may be embodied in some kind of terrestrial conveyor belt (known as the plastic asthenosphere) came about gradually, when submarine exploration using sophisticated remote-measuring devices proved the sea floor was much younger than the continents, probably no older than 200 million years, compared with the continental rocks' almost four billion years. Magma constantly rising from the mantle widens and spreads, and renews the seabed along mid-oceanic rifts, driving the plates on the surface of the planet, pushing the continents along with it. Clearly a balance is maintained where new oceanic crust pushes up old crust in folded mountain chains both above and below sea level.

The truth of the drift argument was clinched after measuring the linear variations in magnetism of the basaltic volcanic rocks along oceanic ridges away from which the sea floor slopes in both directions. The magnetic field variations in the rocks of the sea floor matched up with known polarity levels.[17] There was only one conclusion: as the hot lava oozes out of the oceanic ridges it is cooled by the sea, and takes on the magnetic field of the time. Jeremy Bloxham of Harvard says the field has grown by more than 10 per cent in the last 150 years — a rate of movement, he says, that could mean another reversal in about 1500 years' time.[18]

The leading lights behind the idea of plate tectonics were Harry Hess of Princeton University, and Cambridge geophysicists Dan Mackenzie and Edward Bullard in the 1960s. Mackenzie and others showed that the Earth did not just have a liquid lava-like subsurface. The properties of matter were more complex, and even solids, it seemed, could flow as a result of the tremendous heat emanating from the innards. The Earth is solid beneath its surface, highly compressible yet malleable. It yields to stresses, and sinks and rises accordingly, and it is subject to erosion and deformation. The crust of the Earth splits along mid-oceanic ridges, the mantle wells up to the surface, cools and re-melts, and wells up again before crystallizing into basalt, with the continental crust increasing in thickness from below and in spatial terms. It becomes lighter on its upward journey.

Hence, as some parts of the crust creep forward relative to other parts, stresses build up until the ground literally snaps apart. Fracture planes appear, called fault zones, prone to burst into activity. The energy released by earthquakes can shift thousands of cubic kilometres along horizontal lines of slippage. After the quake the surfaces of the land segments, as viewed from the air, can be seen to offset each other with rows of trees or patchworks of fields no longer in the same alignment as before.

It is now known that Earth's present surface can be divided into about eight major plates, and a dozen lesser ones. In fact, drilling rigs operating from research vessels have produced evidence from igneous rocks that the ocean floor is still being rent by volcanism. And along the mid-oceanic rifts volcanic activity is accompanied by quakes, especially around both sides of the Pacific. They form the infamous 'ring of fire' along the Pacific plate, the Coco plate (westward of central America), and the Nazca plate (westward of the Peruvian coast), and extend as far down as the Southern Ocean. The rate of spreading of the oceans would have taken the present Atlantic Ocean anywhere between 36 and 288 million years to attain its present width. This is roughly in line with the 200 million years estimated for the continents to reach their present positions since the break-up of the super-continent Pangaea.

What is Nature's Message?

Talk of extremes of Earth temperature is bringing us yet closer to a workable theory of extinction. Further ground-work will need to be done in this chapter to uncover the foundations of climate change, and a more detailed examination will be given to climate in the following chapter. At this stage of the argument, however, we must take stock and ask what it is that the Earth is telling us, before we can formulate theories of how our planet has brought about the death of animal and plant life.

Unfortunately, for a variety of reasons connected with the type of evidence that is available, Nature's message is not so easily discovered. There are three reasons for this. The first is when a patiently acquired and voluminous body of knowledge looks less impressive when the suspicion is entertained that Nature might be playing tricks on us. Rocks are sensitive to, and can reflect, environmental conditions. Geological history is contained in strata stretching back more than 380 million years, and is evidenced by the sands, rock, sediment, fossils and volcanic lava. But strata 'records' not only times pertaining to the deposition of substances, but also those times when erosion and transportation somewhat altered the original Earth materials laid down. There are many exhaustive interactions between a variety of processes, many driven by the Sun or by plate tectonics.[19] The record of seas and of temperature of seas (which can give us some clue as to the prevailing atmospheric temperature), is based largely on modern geochemical interpretations of what scientists have dug up and analyzed. Scientists, for example, know the kind of creatures the seas of both the present and the past contain. They know what the fossil-remains of sea creatures look like, and they are aware of the creatures likely to survive best in sea-water of different seasonal temperatures.

The carbon cycle is surprising proof of the organic origin of a great deal of the solid earth that is essentially inorganic. Chalk cliffs and reefs are typical examples. The rudimentary clams were extremely diverse bivalves, looking somewhat like oysters. They built up the reef like the corals found today in the tropics[20] over millions of years, through the accumulation of shells. When the sea level periodically rose many of them died off, but others replaced them when the water became shallow again. Over 50 million years they grew to formations of several hundred metres thick. Their fossils, especially those from the latest Cretaceous, are counted in billions.[21] A litre of sea-water would probably contain upwards of a million invisibly small 'shellfish'. Nannoplankton when they die out cause the deposition of calcareous (carbon-like) substances. Foraminifera, accumulating in large quantities on the seabed, ultimately solidify into chalk. (Whereas foraminifera are one-celled animals, nannoplankton are one-celled plants.) A harder variety of chalk, limestone, arises when the fossils have been thrust above sea level, and thus lack the essential moisture to turn it into 'chalk ooze'.

Important information is also gleaned from carbon isotopes. The lighter isotope (carbon 12) diffuses in solution faster than a heavier one (like carbon 13). During periods of low sea level the rate of organic productivity in oceans should be high, because more nutrients in the form of phosphorus and nitrogen, eroded from the continental crust and washed down to the sea, will be available when more crust is exposed.

Occasionally, however, there seems to be a crisis in reef communities with fundamental changes in the nature and distribution of organic life taking place, and for which there is no adequate explanation. The Paleozoic era presents such a problem. The corals had reached a critical point between the Devonian and Carboniferous, while sponges, bryozoans and brachiopods, which spent their lives rooted to the spot like plants for 250 million years, became more important. However, the Permian extinction seems to have hit these sessile animals the hardest, with the much rarer mobile organisms like bivalve molluscs surviving relatively unscathed. Nowadays the shrimps and molluscs are much more common than the brachiopods, which seem never to have fully recovered from their traumatic routing.

The second reason why Nature's message is not so easily deciphered is because the fossil record is biased towards the preservation of marine life in shallow water. Precisely because sea level varies so much the geological record is often hard to interpret. During transgressions, with the sea coming in, marine sediments are deposited on the top of land sediment. During regressions, the high ground is eroded and the record is often obliterated.[22] And when the sea comes back again marine deposits are dumped over wiped-out continental deposits. The best place for a permanent record is thus the ocean floor, and such deposits are usually got from deep-sea drilling vessels.

This means that fossils from the interior of continents where most species develop, and where there is a much wider range of habitats, are largely absent, so it is difficult to know whether species stopped evolving in continental interiors. Edwin H. Colbert said that if extinction was caused by a change in the food supply, it was too subtle to be registered in the fossil record.[23] Robert Bakker says circumstantial fossil evidence is flawed because of the difficulty in separating relevant from irrelevant details. Depositional conditions in many cases were not suitable for the preservation of very small skeletons, like those of lizards and snakes. The fossils also form larger natural barriers so that species eventually become new species, what Bakker calls a 'high topographical-geographic diversity'.[24]

The third reason is that fossils are often too sparse to enable demarcation lines to be clearly drawn, as at the Cambrian/Precambrian boundary. Some fossils have no body as such, and are identified by the tracks or trails they leave, or by their borings, faeces and burrows, etc., which can leave plenty of room for error. The traces of soft-bodied creatures are seldom preserved, and one type of organism may produce different traces; many kinds of organisms may produce the same kind of trace.[25]

Even defining 'mass extinction' is not easy. The chronological series of any species is known as lineage. But such linguistic definitions can be a substitute for *de facto* physical earth changes. By evoking the concept of lineage, one can sidestep the notion of extinction by converting older species into new ones. Real extinctions are when species fail to transmutate into different forms of themselves. This is especially apparent when the ecological niches they leave behind are occupied by totally new species, such as the giant reptilian dinosaurs by smaller and physiologically different mammals.[26]

The number of species alive at any one time is known as a diversity, and useful comparisons can be made between diversities so that we can get some fundamental idea of the measure of change over geologic time. The problem is that periods and epochs differ greatly in duration (i.e. the Ordovician period is twice as long as the Silurian), so naturally one would discover more extinctions, in terms of straightforward body counts, in one epoch or period than another.

Then there is the 'background radiation' which occurs naturally. Half of any total of mollusc species become extinct virtually every seven million years. Species are fairly low down in the scale of things, below families and orders. If families or whole orders are chosen for analysis the results can be misleading. If you choose classes only for extinction, not even the dinosaurs came to an end at the K-T because *Reptilia* still survived.

Drift — An Intermediate Explanation?

Whatever the difficulties associated with interpreting fossil evidence (and placing it within accurate temperature zones) the strata evidence of sideways movement is a useful and important clue. For we have definite proof of drifting land masses, and hence of changing sea levels, and ultimately of changes in climate. We shall explain more precisely how this could happen in the following chapter. First, let us return to a discussion of the seas, since the extent and height of sea levels has played a major role in the survival or otherwise of species.

It is clear now that the drifting continents, over millennia, brought about crucial geomorphological changes that shaped the planet. There is much scientific evidence to show that biological life has flourished in geologic eras when land masses were spread more equidistantly, and when sea levels were neither too high nor too low, such as in the Carboniferous about 340 million years ago. Sea levels were possibly rising everywhere in the Cretaceous period to become moderate in the Cenozoic and Quaternary period. This is where climate, as a result of geophysical changes, emerges for the first time as an integrated explanation. For scientists believe a steady-state climate began to disappear not only when the major land masses assumed a certain configuration, but when distinct atmospheric changes began to occur for the first time, and when it is traditionally believed oxygen reached about 18 per cent of the total.

Evolutionary processes forged ahead. The Triassic appears in retrospect to be marked by the gradual extinction of groups of tetrapods and sundry vertebrata originating in the Permian after the original amphibious tetrapods, while other vertebrate groups gradually came into being — turtles, crocodilians, flying reptiles and mammal-reptiles. By the Upper Triassic early reptiles had reached their widest distribution, and most varied development. By the close, primitive reptiles and amphibians dropped out, to be replaced by early dinosaurs, probably due to evolutionary competition.

Life in the sea flourished; there was a rich variety of molluscs and ammonoids, but of course many fewer brachiopods. Among the coral barrier reefs could be found the placodonts and northosaurs and tortoises, all preying on various sea plants and seahorse-shaped cephalopods. At the end of the Triassic, fish lizards all died out. Extinction at the end of the Jurassic also eliminated many of the dinosaurs; probably, according to the fossil record, there were as many as eight sudden mass extinctions among the dominant vertebrate families.[27]

As time passed North America, while still attached to the southern continent by the isthmus, seemed to curl outwards and northwards to put first several hundred then thousands of miles of ocean in between the respective coastlines. The South Atlantic emerged, with South America putting ever more distance between its coastlines and that of Africa. India, already separated from Antarctica (itself a much larger and somewhat more northerly continent than it is today), headed further north to eventually join up with Asia. Australia has 'floated' over hundreds of millions of years, on a sea of upwelling magma, towards the South Sea Islands of the Pacific until today they are almost geologically touching. Africa and Eurasia were much wider apart than they are now, and the Atlantic was narrower. But because of the width of the Tethys (an oceanic gulf on the eastern side of Pangaea) the current was more of a clockwise eddy than a westward-flowing current like the Pacific, as the fossil evidence seems to show.[28]

We can now attempt to arrive at a theory that will help us better to understand both evolution and extinction. Throughout history, for most of the time, parts of the sea intercommunicated but the land, as today, did not. So land fauna are more likely to lend themselves to regional evolution than the fauna of the seas. We can imagine, for example, a species trapped on an island with an increasingly arid climate, with random mutations producing creatures better able to withstand climatic rigours.

Expansion and contraction led to profound ecological disturbances amongst offshore and lowland families, with the world ocean currents moving vertically and high biological innovation occurring without much speciation.

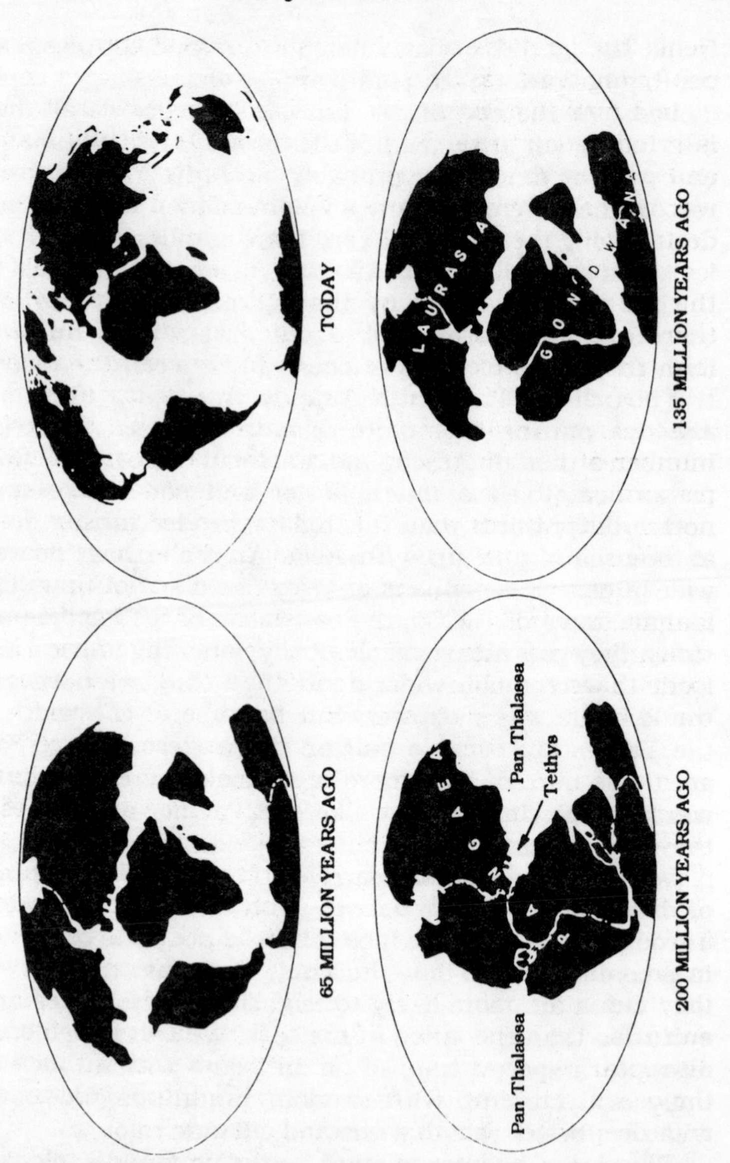

Fig.23. Continental drift: some 200 million years ago Pangaea began breaking up. (Source: Preston Cloud: Cosmos, Earth and Man.*)*

Hence the land/climate relationship is an important one, conferring great explanatory power.

In a further example, we can imagine a continent that is grinding northwards, with the climate getting cooler and putting an ever-widening ocean between it and the warmer southern shores. Global cooling naturally destroys life in what would have hitherto been the 'tropics', as it has nowhere else to migrate to. Only species in the non-tropical zones can migrate to warmer areas where there are no barriers, or, in the case of a warming (brought about by other events), to more northerly climes.

The nature of the terrain also would have greatly affected co-evolutionary developments, and the total number of vertebrates on land must have been directly proportional to the amount of land available. There would not, in other words, have been many vertebrates on land at the beginning of the Jurassic, since the seas were more widespread. Deserts remain effective barriers between faunal 'provinces' to most land animals, and the ocean similarly represents a natural barrier.[29] Given the right food, the right climate and absence of danger, species would migrate ever onwards by several feet per year.

Clearly, then, those creatures that are most vulnerable are those that are not distributed widely enough. Even poorly adapted species have a better chance of survival if they are dotted about continental areas in the right places.[30] Robert Bakker goes along with this drift theory of extinction. He refers to the insidious and strangely unecological factors involved in the way moving land masses can disturb the habitats of living beings, the way they can attack the 'evenness of an ecosystem'.[31] Great animals, he says, die off when their favourite haunts disappear, when the warmish surface waters which kept the ocean temperatures even are more thoroughly mixed with deep water draining off from shallow seas.

This latter event can have serious consequences. At lower depths the waters can become anaerobic, or stagnant, too deep and unstirred to absorb surface oxygen. In the basins where water levels were going down the level of oxygen was becoming depleted, causing crowding together and epidemics, with animal carcases atrophying. This can also explain why marine organisms died off. In contrast, deep oceans can compensate for this by supplying cold, dense water — descending from the surface. In the past this paved the way for primitive oxygen-breathing creatures to come ashore. It also explains why shallow marine inhabitants are often the catalysts for evolutionary change. We have seen this in the Devonian period, and with the emergence of the first amphibians.

Throughout their existence as a species the dinosaurs were on the march — they 'radiated' across the globe in two different ways. One way was simply to stroll over large interconnected land-masses to dominate life from the Jurassic until the dawn of the Cenozoic. But dinosaurs also 'radiated' in the conveyor-belt sense: for example, diplodocuses, brachiosaurs and iguanodons were found in Africa, and in the late Jurassic in northern America, when broad lowlands existed in many parts of the world, many carnivores, theropods, sauropods and stegosaurs thrived. It is, incidentally, the period when the first flowering plants appeared. We cannot help noticing that the various species of Mesozoic dinosaurs became extinct when the land masses began to approach the positions we know today. As a result the Jurassic and the Cretaceous climates must have been becoming similar to those of today.[32]

However, there is yet another caveat: we must be certain that the direction and time-scale of drift is right, and that we are not talking about drift after the event. Drift at a few centimetres each year sent some continents to quite high latitudes — hence dinosaur footprints on the island of Spitzbergen. But we know that they were carried there long after the dinosaurs themselves vanished.[33] On the other hand, Patricia Rich and her colleagues from Monash University, Australia, in an article in *Science* magazine, described dinosaur fossil finds from an emerging Australian sub-continent at the end of the Mesozoic that was actually still within the Antarctic Circle. Antarctica in early Mesozoic times was more northerly than it is now, but it drifted further south although the gap between it and Australia was still narrow. The separation created broad rift valleys for plants and animals to colonize, valleys that later opened up to form part of the Southern Ocean.

Some of the dinosaurs identified by the Monash researchers have been found nowhere else, hinting that Australia was zoologically isolated from the rest of the world. So a cooler Antarctic/Australia supported not only dinosaurs but stands of conifers, and ferns, plus a variety of insects and other invertebrates, shrimp, fish, amphibians, turtles and pterodactyls. There were at least three kinds of small theropods called hypsilophodontids, which were preyed on by carnivorous dinosaurs. One of the hypsilophodontids had an unusually large brain and pair of eyes, adapted to the poor polar light conditions where the Sun remained below the horizon for as long as two months every winter — proof enough that they could hardly have been pure poikilotherms.[3 4]

Chapter 9
THE CHANGING MESOZOIC CLIMATE

THE PHYSICAL principles of continental drift and extinctions go some way to explain how co-evolution, in a sense, gives way to co-extinction. But we need to know a lot more: to consolidate the argument, and to introduce specific mechanisms.

We have already seen that when land masses drift apart great ecological and geophysical events take place. Climate — meaning enormous long-term changes in atmospheric pressure and temperature — not only emerges as a phenomenon for the first time, but climate gradually changes *as a function* of continental drift. We could therefore postulate, in loose terms, that continental drift, *via* 'climatic change', was responsible for the final demise of the remaining ceratopsians. But before we could begin to spell out in detail how this might have happened, we would be confronted with explanatory problems of a more conceptual kind, for we cannot really assert that climatic factors played a crucial role in dinosaur extinction if we have yet to decide whether we are talking about tens of years or millions of years.

One insuperable problem is that our description of proven facts about the location of continents masks a margin of error spanning a period of 100 million years — a very broad yardstick indeed. Similar handicaps arise when theorizing about a catastrophic end to the Cretaceous period: few meaningful timespans can be discerned from the stratigraphic record. This is because the rock strata are used, often tautologically, to demarcate the end of eras and the end of epochs as a matter of definition. Fossil remains show extinctions worldwide, both continental and marine, but the width of any band of minerals or fossilized sediment, measured in, say, centimetres and used as a yardstick to demarcate eras, is not easily translated into human time frames.

The top layer of chalk and limestone at the Maastrichtian overlays a thin layer of grey clay with fish fossils embedded therein. But clay is non-organic in origin, made up of fine dust from continents, or even from space; and fossils in the upper layer of clay and limestone are totally different from the ones below.[1] At most, geophysicists might be able to reduce the timespan of the deposition of the various clays and chalks to a few tens of thousands of years, but not to a shorter period. We need, then, to sort out the various types of geophysical evidence, because the clues that are yielded up vary in the accuracy with which they can reflect human time frames.

At least we know that drift brings about an eventual change in the weather triggered by the emergence of long-term and gradualist changes in Earth temperatures. Weather is determined, in the main, by different heat contrasts between the equator and the poles. It is when large longitudinal oceans (going up and down the globe, vertically) rise when land masses break up, that temperature contrasts emerge. When the seas flood continental margins, as they did in the middle Ordovician and the late Cretaceous, this must have a marked effect on long-term weather patterns. Scientists can work out the rate of precipitation in the abstract: if you have a contrast of, say, 8°C (22°C for oceans, 30°C for continents), then you get a rainfall of 53 inches and a steady wind speed can be achieved. The heat released by condensation amounts to a warming of the atmosphere of nearly 1°C with some of it returning to the maritime atmosphere.

But there is another complication arising from growing and vanishing ice-packs and continental drift. An important clue to climate formation is related theoretically to drift: it is the fact that the ratio of the sea to land has constantly changed throughout time. It was previously thought that drift would only partially explain this; now scientists are beginning to believe that drift can explain *everything*, including evolution, changing climates and the extinction of animal and plant species.

A traditional, well established theory for sea-level change focuses upon what happens when the Earth's considerable ice packs wax and wane, for at least thirty major, and hundreds of minor, oscillations in sea level have occurred in the past 600 million years as a result of eustatic change. (*Eustasy* means the global rise and fall of ocean basins, and other complex theories relating to gravity, Earth density, the speed of Earth's rotation, tidal forces, and so on.) However, a major new terrestrial theory to account for the Ice Ages appeared in 1989, since climate researchers have long recognized that the earlier Milankovitch theory has difficulty explaining some of the Ice Age cycles. Let us try to see why.

Milankovitch's cosmic theory was both interesting and plausible, since it extrapolated from what we know about the cause of seasonal variations in heat and cold. Because of the tilt of the Earth, the northern hemisphere gets fractionally more solar heat at the perihelion (the Sun's closest approach to Earth) after the Winter Solstice on 2nd January. This means that northern winters are milder than they would be if the orbit was entirely circular, but the northern summers are commensurately cooler because six months later the Sun is at its aphelion (furthest point away from the Earth). However, the fact that the Earth's orbit changes gradually as time passes inspired a Serbian mathematician, Milutin Milankovitch, to develop a theory, which he formulated between the two World Wars, to explain Earth's periodic Ice Ages.

Milankovitch also reminded us that the direction of the axis changes, too, over a period of some 21,000 years due to the pull of the moon, so that halfway through this period the summer and winter solstices are reversed and the northern summer will be commensurately hotter. The tilt of the axis also varies over a further period of 25,000 years. Furthermore, the Earth's orbit gets progressively more, and then less, eccentric over a round period of 100,000 years. If all these variations are taken into account, said Milankovitch, then over 100,000 years there would be a tendency for mild seasons to be interspersed with extreme seasons — say, roughly over a duration of 25,000 years each, with the 100,000-year cycle peak bringing about an extra degree of severity. It is these orbital, or cosmic, periodicities that bring about Ice Ages and inter-glacials.

Nevertheless, it has recently been pointed out that the 100,000-year cycle only changes the amount of solar radiation received by less than half a percent. In contrast, the shorter cycles of Earth wobble-and-tilt can have as much as a 20 per cent effect. The problem of why the northern and southern hemispheres need to be globally synchronized (with the same kind of knock-on effect), when the orbital calculations show that the southern hemisphere should be getting more sunlight during a northern Ice Age, was also not satisfactorily explained.

Instead Wallace Broecker, a distinguished geochemist at Columbia University's Lamont-Doherty Geological Observatory, said that Ice Ages can be internally created, with the Earth able to switch back and forth between the two modes almost instantaneously. This is because the feedback loops governing climate are probably more pronounced than is usually thought. The key, says Broecker, is the North Atlantic, and the amount of meltwater draining off into the warm Gulf of Mexico and later from the more northerly Great Lawrence Seaway. He takes as read, as do many scientists, some orbital factor responsible for getting an Ice Age started. After that, however, the terrestrial cycles take over. The warm Gulf Stream gets jammed, cooling the north still further. When the warming gets going as the meltwater flows down to the Gulf, the more northerly St Lawrence Seaway finally becomes ice-free but the more frigid fresh water (which does not sink) just piles up and causes a temporary cold reversal.[2] This accounts, says Broecker, for the frequent mini Ice Ages that seem to occur for no orbital reason, such as that which occurred some 3000 years ago.

Broecker's theory is similar to that publicised recently by a Leningrad-based oceanography research group attached to the Soviet Academy of Sciences. Their computer models show that, regardless of external factors, the Earth's climate extremes occur as a result of the changing distribution of land and sea over tens of thousands of years, and particularly, to take one example, when a landlocked ocean is at one end of the poles.[3]

The Soviets were elaborating on a theme that has been well known to Earth scientists for a long time. Earth's metabolic complexity is probably unique in the universe, but especially so because of the arrangement of its terrestrial land masses and surface water. This arrangement is untypical of the other rocky planets such as Mars, Venus and Mercury, even the moon. Although it is known that Mars and Venus are, or have been, volcanically active, it is unlikely they would possess moving continents driven by internal heat.

The task of the geology teacher would be made a lot easier if Earth was bone dry like Mars, or instead was covered with a uniform envelope of fluid. Unfortunately, as we have seen in Chapter 1, if this were the case intelligent life would probably never have arisen; indeed, there would have been no animal life of any sort. At present 71 per cent of the globe is covered with water, although 81 per cent of the southern hemisphere is covered by seas and oceans. This has not always been the case. Throughout time the seas have changed place with the land which would alternate between becoming parched or flooded.

Because of this we can make educated guesses about what the climate might have been like in the past simply by imagining the likely distribution of land and sea that would have been possible. Clearly, as water is a good retainer of heat, global temperatures would be considerably higher than they are now if there was less land on Earth. The energy that enters the Earth's surface is balanced by the complicated way heat is retained or radiated outwards. Heat is also transformed into the kinetic energy that moves air, which drives windmills and distributes atmospheric heat across the globe.

The volume of the ocean basins has fluctuated over time to act as a great life-shaping physical force. Since oceans take a long time to warm up and a long time to cool down, they have a moderating influence on climate. It follows that when ocean levels are lower, cooler temperatures are likely to occur, being often both the initiating agent in an Ice Age and the chief mechanism for prolonging it. Extensive plateaux of dry land tend to get hot during the day but cold during the night, as heat escapes back into space. Temperatures would reach an optimum 12°C higher if all the present land masses were concentrated around the equator, with oceans to the north and south. Something like this probably happened between the Triassic and the K-T boundary. Both polar regions were oceanic, and this is thought likely to favour an equable world climate.

In an opposite hypothetical scenario, with the land masses being much more extensive to the north and south and the world's oceans somehow located largely around the equator, the Earth would be locked into a permanent Ice Age of unparalleled severity.

It is already possible for glaciologists to work out how long Ice Ages lasted. Unfortunately scientists are in dispute as to whether Ice Ages — by themselves — in the distant past caused extinctions. True, during the late Precambrian, according to the evidence of ancient sedimentary deposits, a 50-million-year Ice Age lasting into the Silurian put paid to a whole host of species.[4] On the other hand, another Ice Age extended from the late Carboniferous up to the end of the Paleozoic, a period spanning 90 million years,[5] a period when there were no extinctions.[6] And no other major Ice Ages occurred until about 22 million years ago, in the early Miocene,[7] measured by a marked decline in sea temperatures during the Oligocene and Miocene periods. Further, we can almost certainly rule out Ice Ages as a cause of mass extinctions in recent historical times. There is no convincing evidence of major polar ice caps at the end phases of the Permian, Triassic or Cretaceous, let alone large-scale glaciations. If our facts are right, then, this means we can rule out Ice Ages occurring after 100 million years ago.

Drift and Climate

Broecker's theory, and the Soviet theory, in the meantime emphasize terrestrial features in global changes of temperature, and focus our attention on the changing size, depth, density and ultimately the temperature of the oceans.

An excellent illustration of how drift causes rises and falls both in sea level and in land contours, comes from Damian Nance and Thomas Worsley of Ohio University. They suggest that the sea floor is created by the upwelling of hot material as the mantle is pushed apart by the spreading sea floor, and continental land masses are thrust upwards. In the process the floor becomes denser and sinks, and with it the ocean above. Hence sea level would fall over time if other compensatory factors did not come into play. Many scientists believe that the mid-oceanic ridges are also created as the sea floor spreads, displacing ocean water and making sea levels appear to rise.

Clearly something like this must have occurred very early on in Earth's natural history: climate, we reiterate, is *largely* a function of drift, with solar output or orbital tilt models playing either a subsidiary or initiating role.

Let us pursue the logic of this perspective by backtracking to a momentous event: the day the world's first super-continent was created. From then on, sea level could no longer remain constant because it would be dependent upon the changing configuration of the surrounding land masses. By the middle of the Permian period the sea separating the two halves of Laurasia had virtually been squeezed out of existence, to have harshly

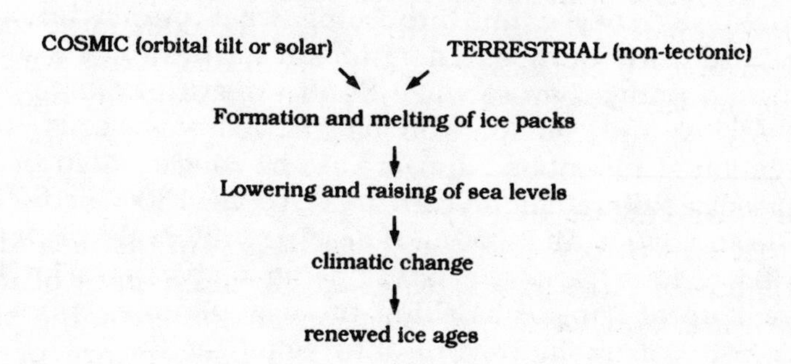

REVISIONIST ICE-AGE THEORIES

plate tectonic activity

↓

rifting land masses

↓

new mountains and ridges

↓

fluctuating sea levels

↓

fluctuating climate

↓

ice ages

Fig.24. Most climatic change theories focus upon the way Ice Ages come and go, and this illustration shows the reasoning behind different schools of thought.

dramatic consequences. Just how dramatic we can illustrate by pointing to any familiar mountain range.

If we were to describe the genesis of the Alps, we would say they came into being when Africa literally collided with Europe. The Himalayas, the Rockies, the Andean Ranges, arose when South America became part of Africa, and the Amazon had flowed westwards. The creation of mountains appeared to be concentrated in six episodes; 2600 million years ago, 2100m, 1800m, 1600m, 1100m, 650m and 250m. These are intervals of about 360 million years, the latest being responsible for the break-up of Pangaea. Behind the new mountains a wide diversity of habitats came into being for the first time: deserts, tundra, and, heading down further towards the coastline, marshlands and rivers. Where the continent ends we find shallow, warm, continental shelves suitable for marine reptiles. Oceanic ridges play their role in the saga, brought into being by the same crushing forces; they are in a sense oceanic versions of mountains.

When drift slows the opposite happens: the mid-oceanic ridges become slowly worn away, inland mountains start to erode, the habitats that once provided self-contained niches begin to disappear. When drift is at a complete standstill, the sea drops even further as the mid-oceanic ridges become virtually non-existent (because the sea is taking up the space of what were massive underwater mountains). There is natural drainage as water flows away from the coastal margins with all the inevitable ecological consequences.

Here we must pause and remind ourselves of what is going on. Since the beginning of time drift has pushed continents both away from, and towards each other. Often this has happened at a rate of two inches a year, as with the present drift of America away from Europe. At the risk of over-simplification we could say that the former event causes a warming, and the latter a cooling over large areas of the Earth where the most marked of these events is taking place. The former, for example, happened during the break-up of large land masses between the Permian, some 280 million years ago, and the Eocene, 55 million years ago. Then more of the Earth's surface becomes subject to the mild oceanic influences of the gulf-stream kind.

As we have seen, as drift proceeds sea levels *tend* to fall, but they can be checked and sometimes reversed according to the density of the ocean floor and the extent of underwater displacement ridges. Even so, large continents, as today, are still frigid for long periods of the year. Inland they can become cold and arid, with little seasonal change; or there can be dramatically contrasting change, with sparse vegetation suitable only for very small mammalian and amphibious creatures. But temperatures can be greatly eased by inland waterways and seas. The main causes of cooling when a coastal transgression takes place are the location of mountain ridges and the draining away of inland waterways, restricting the circulation of warm ocean water, and leading to the build-up of ice at high latitudes. When this happens habitats shrink in size and become less varied. Primary producers would have declined in number, and the intricate and interdependent food web become seriously disturbed.

As a general rule the wetter the land surfaces, and the moister the air of those small land surfaces surrounded by equable seas, the higher the atmospheric temperature. Sometimes the proximity of land to water is too intimate, and even a moderate rise in sea level can be enough to flood large areas.

It is with these geophysical facts in mind that scientists explain the Permo-Triassic extinctions of such small creatures as the trilobites and graptolites, the main species alive at the time.[8] Regression led to an increase in competition on continental shelves, the result of which was to devastate coral reefs and organisms associated with them. Thomas J. Schopf advances a rise-and-fall drift explanation to account for what was going on: the shallow marine seas of the early Permian were reduced from 40 per cent of their possible distribution to less than 15 per cent in the later period.[9]

The later Permian regression could also have been due to water withdrawing into a deepening ocean basin which continued to sink due to climatic cooling, as well as to the effect of spreading ocean ridges.[10] This got rid of labyrinthodont amphibians and most mammal-reptiles in the process. About 20 per cent of all marine invertebrates also disappeared.[11] The ammonoids suffered, as did the brachiopods and gastropod molluscs, with only the ichthyosaurs surviving into the succeeding Triassic period. Benthic foraminifera survived, as did many types of lung-fish, but a great many aquatic animals died out.

The Record of the Seas

We have so far been dealing with the most approximate yardstick for assessing the speed at which extinctions occur. It is time to move on to the Cretaceous extinctions, and to look at more accurate tools of measurement.

The Cretaceous, too, had widespread shallow seas, especially across the west and south of northern America.[12] The period in the early stages was a time of transgression, to be followed in the later stages by regression.[13] This implies there was first a warming, and then a cooling, a point confirmed by other scientists as we shall see later.

Here we can turn to the measurement of ancient sea temperatures themselves; in fact, we can improve our time-scale by a full factor of ten. Oceans offer the most important clues as to how species fare and *where* they fare.[14] It was always the organisms adapted to warmer seas that were especially vulnerable to any oceanic cooling, as paleontologist Steven M. Stanley proved by pointing up the decimators of marine species in the most recent Ice Age, with only those able to migrate to warmer zones (such as fauna in the Pacific) being spared. Similarly with the Late Eocene extinctions, it was the trapped species in the warm coastal areas that suffered most.

Similar changes in sea temperatures may explain many other extinctions in the distant past. Patrick

Brenchley and his colleagues at the University of Liverpool report on rock findings from the Hirnantian glaciation some 439 million years ago in the Ordovician period, when the Earth cooled into an Ice Age. Brenchley found three distinct phases of extinction with, first, the trilobites and echnoderms dying off in the temperate zones; then the inhabitants of shallow tropical seas died off; the last to become extinct were the brachiopods, corals and bryozoans. This is proof, he said, that climatic change rather than impacts may have caused past extinctions. First there was a cooling, then sea levels fell as the water froze, and finally the seas rose again as the climate warmed. In actual fact, while the temperate zones shrank remorselessly, a brachiopod that preferred cooler water came to dominate the environment in the glaciation.[15]

It is not just ocean warmth but the quality of seawater that counts. Sometimes reduced salinity at the surface and a severe oxygen depletion beneath the surface can bring about the demise of much marine biota. This can happen when oceans are covered by a layer of low density water from a brackish arctic ocean.[16] Turtles and crocodiles are the least likely to have been eliminated in this way. Some reptiles could adapt to brackish water, provided the food chain remained relatively intact. Those living in moist coastal areas would have benefitted, too, as would others living in environments of the salt-marsh types. And this despite such habitats being too limited to support a community of large dinosaurs.

Broadly speaking, there has been a discernible decline in sea temperatures over millions of years, as the continents have continued their more or less steady drift apart as the result of mountain building and crustal movements.[17] H.C. Urey, the scientist who gained fame as the first man to create 'life' in a test-tube (*see* Chapter 1), and his colleagues have invented a 'thermometer' based on oxygen isotope studies (where two different varieties of atom vary with a change in the temperature) in sea fossils to measure ancient ocean temperatures.

By 1950 Urey and Cesare Emiliani, who was later to

become an eminent climatologist, had honed the techniques so well they could tell in which season a creature was born and in which it died off, and for how long it lived.[18] And, more significantly, foraminifera fossils suggest average temperatures of the sea at mid latitudes seemed to decrease by about 5°C (9°F) during the closing stages of the Cretaceous, declining from equatorial/polar temperature differences in the sea of 15°C earlier on. Since the 1950s other laboratory experiments suggest that the atmospheric circulation is dominated by thermally driven cellular convection; by rising motion over warm areas, and sinking over cool ones. Without the equator/polar contrast, much circulation change would have been determined by extensive land-sea breezes on coastal areas, and seasonal climates would have been virtually non-existent. Although measurement of ancient sea temperatures can tighten up our knowledge of climatic time-scales, some scientists believe sea fossil evidence can work very well in tandem with a knowledge of drift.

These days powerful computers are employed routinely to assimilate large quantities of information in regard to ocean currents, wind speeds and salinity in order better to understand the ocean-atmosphere interaction. Scientists at the Earth System Science Centre in Pennsylvania University in 1989 were using new computer-generated maps which can work out the patterns of ocean circulation as they probably existed some 100 million years ago. There is also the geological evidence which needs to be matched up. Erick J. Barron and William H. Peterman of Pennsylvania State, using their computer, chose to study the Cretaceous because the geography of the period is fairly well known, and other computer attempts have already been made with which to make comparisons.[19] Even with this streamlined and computerized approach there is still a limit to human knowledge. It is, for example, difficult to work out the atmospheric circulation patterns of 65 million years ago, because of the differences in continental positions and the height of mountains that existed at that time. This is why many scientists look at what seemed to be happening to the world's stock of vegetation for yet further clues.

The Evidence of Plants

The fossil record of changing flora is thought to be particularly useful because Mesozoic plant life can be compared with what is known about modern plants — and the 'life zones' in which they flourish can give a knowledge of climatic changes down to about a million years. Flora exists above high strata where dinosaur bones have been discovered: in Alberta up to six metres higher. Much of this is compatible with the fossil record of vertebrates, and points to a small latitudinal temperature contrast.[20] Tropical vegetation extended 45 to 50°N of the equator, and south of it. In western North America tropical and subtropical floras were replaced by floras indicative of temperate or seasonal climes, or both.

Some typical Mesozoic plants like cycads, say some scientists, favoured high stable temperatures[21] such as those found nowadays in New Zealand, South Africa and South America, with conifers and horsetails pushed into a minor role.[22] This, however, would be difficult to prove. The problem with gymnosperms (or naked seed plants) is that they can either conjure up images of palm trees (like the cycads) or pine trees (like the coniferals of the Permian and Triassic periods). One would, after all, expect gymnosperms to dominate in a cooler climate, as they do today in the northern hemisphere. Today's natural conifer forests tend to exist in marginal areas with poor soil, poor climate or high latitudes. They are mainly today found in Northern Europe and Scandinavia. As a general rule, where temperatures fall below minus 50°C it is a certainty that only conifer forests exist.

This is where the controversy about a global cooling at the K-T fits uneasily into the picture. The consensus of opinion is that the rise of the angiosperms was facilitated, if not directly caused, by a change to a cooler regime, and this is said to be evidenced by the fossil record of plants.[23] There is, of course, a danger of circular reasoning here ('a cooler climate changed the vegetation; therefore changed vegetation is proof of a cooler climate'); coincidence could play a much larger role than is thought. Nevertheless the arguments for climatic change are persuasive. Smooth-margined leaves did replace tooth-margined types typical of deciduous hardwoods,[24] and there was a replacement of sub-tropical flora by temperate flora.

We have already suggested in this book that the rise of the angiosperms was sudden in geologic terms, but still not abrupt enough to cause animal extinctions, say of those herbivore species used to gymnosperms at the beginning. Change, it has been said, was 'geologically rapid but ecologically slow',[25] with the transition to a cooler climate taking about 12 million years. Kenneth Hsu goes further and says that unpalatable vegetation, which in the early first few million years was probably confined to upland areas, became more commonplace. The shedding of winter leaves (which the gymnosperms did not do) also led to possible scarcities.[26]

Recently, co-evolution (or rather co-extinction) has taken on a more direct meaning. New work on plant evolution suggests that the thorn and the angiosperms may have been encouraged to grow with the help of pollinating bees. More significantly, their radiation may have been speeded up by the dinosaurs literally blundering through coniferous forests of the Cretaceous, leaving a strip of opportunities for the early flowering plants, and unwittingly bringing about their own demise.[27] In any event, as temperate forests moved south they infiltrated the existing fauna, and dinosaurs as a consequence found it less easy to adapt.[28] Nevertheless, it is now becoming clear that the changing nature of Earth's vegetation in the Cretaceous, and the piecemeal nature of species extinction, casts considerable doubt on the asteroid-impact theory. Art Sweet, of the Geological Survey in Canada, and his colleagues have tracked down the evolutionary fate of nearly 300 species of ferns, evergreens and angiosperms — by studying the fossilized spores of plants in Canadian soil — over the period approximately from 70 million to 63 million years ago. During this period there was a gradual decline in species diversity, becoming particularly apparent 300,000 to 400,000 years before the K-T boundary. The Canadians agree with the impact theory up to a point. But whatever extinctions occurred, they were 'superimposed' on an established pattern of dwindling species diversity, and this, they say, was probably triggered by long-term climatic changes.[29]

A final word of warning: climatic theories of vegetation-change lose a lot of impact when we bear in mind that it is not just climate that determines their presence but the speed at which saplings grow. W.J. Bond, of the Department of Biology at the University of California at Los Angeles, advances a 'slow seeding' hypothesis, arguing that they can be literally overshadowed by the spreading canopies of young hardwoods and angiosperms. Angiosperm leaves often adjust their shape and size to suit local conditions. They take advantage of light shining through gaps in the canopy, and the water-carrying vessels in their trunks are more efficient. Climatic change theories could be partly rescued in the knowledge that some angiosperms, being unsuited to the more rigorous climes of the northern hemisphere (unlike the conifers), suffered a lot in winter because their leaves would be continually subject to the risk of frost. The water vessels within the tissue of the plants then tend to accumulate air bubbles, which can block the flow of nutrients.[30]

What Kind of Atmosphere?

Yet another way of working out what happened at the K-T is to look at what was happening to the changing atmosphere. The catastrophic argument advances a variety of theories to explain a *sudden* (in geologic terms) chilling or warming at the very end stages of the Cretaceous. The immediate causes could be due to an increasing opaqueness of the atmosphere, or to forest fires, or to massive injections of carbon dioxide into the seas or atmosphere. Other scenarios include the toxication of what would otherwise be normally life-supporting gases, such as nitrogen and oxygen.

In the meantime an astounding piece of thrown-away information, appearing in *New Scientist* in November 1987, could cast an entirely new light on the K-T mystery. At the same time it demonstrates that there is no scientific consensus about the gases in the atmosphere over vast stretches of prehistoric time. For, according to Gary Landis of the US Geological Survey in Denver, and Robert Berner of Yale University, the Earth's atmosphere only 25 million years ago had an oxygen content that was still 5 per cent less than it is today. But 80 million years ago it contained 50 per cent *more* oxygen than it does now. Landis and Berner said this was an absolute increase; in other words, it was not at the expense of nitrogen, which is a highly stable gas. This denser atmosphere, it was claimed, also helped the ungainly pterosaur (with a wing span of up to eleven metres across) to stay airborne.[31]

Landis, in the course of reporting new American fossil findings, used an instrument called a Quadropole Mass Spectrometer to analyse the gases emitted after crushing an 80-million-year-old fossilized tree resin containing microscopic air bubbles. In anticipation of the argument that the samples could have been contaminated, he said the dense pressure of the atmosphere at the time would have prevented gases entering the bubbles.

Raised oxygen levels could possibly have come from the first generation of flowering plants, but it is difficult to see how they could have *greatly* increased this gas without any supporting evidence from pollen fossils of much higher densities of vegetation on Earth. And we must bear in mind that it was after the Carboniferous period that the flowering plants became dominant. This *could* have caused oxygen in the atmosphere to drop. Such is the conclusion of Jennifer Robinson of Pennsylvania State University, who claimed in 1990 that most trees in the Carboniferous contained much more bark than wood, and bark is rich in lignin. This substance is difficult to degrade with atmospheric oxygen, so the dead trees ended up as vast coal deposits, and the oxygen that would have been used in degrading the wood built up in the atmosphere. She says, using Robert Berner's isotope evidence, that the concentration of oxygen rose steadily from a level of about 15 per cent some 380 million years ago, to reach 35 per cent during the Carboniferous. Then it dropped to about 30 per cent in the Cretaceous (as we have seen from Berner and Landis's evidence) when lignin-rich trees became extinct and were replaced by other kinds of vegetation.[32] Today the oxygen content of the air is around 20 per cent.

We must also remember that high levels of oxygen can be poisonous for some species. Indeed, dramatically raised oxygen levels could well form the basis of a catastrophic explanation. Certainly biological systems accustomed to one level of oxygen could have profoundly affected evolutionary processes. The theory is that the total volume of carbon is reduced as the land is further encroached by plants. This scenario is also a self-limiting one. Too much oxygen will obviously prevent plants from continuing to invigorate themselves with the atmospheric carbon, hence they will eventually die back. On the other hand the meteoritic impact allegedly brought about widespread global fires. But if the oxygen content exceeded 30 per cent for whatever reason, it could itself have caused widespread and spontaneous forest fires.

The interaction between atmospheric gas change and oceans is often the result of historically unusual imbalances in molecular arrangements. Some evolutionary biologists believe raised CO_2 levels could have been caused by falling sea levels and subsequent species over-crowding.[33] This argument is marred by the fact that if ocean micro-organisms die off they take out less

CO_2. At present half of the world's organic carbon is locked up in soil and rock (*see* Chapter 1). It was recently suggested that the Permo-Triassic extinctions were caused when the sea became lowered. This meant that additional organic-containing sediments in turn became exposed, releasing more carbon dioxide into the environment, and disturbing the oxygen-carbon balance in the process.[34] A variation on this theme comes from Erle Kauffman of the Smithsonian Institute, Washington. He prefers gradualist explanations involving oxygen depletion in the ocean related to climate change.[35] This can also be complicated by the reduced density of ocean water when it has been infused with too much clear water. Temperatures would then start to rise again. Normally, however, the plankton would absorb any increase in atmospheric carbon, so once again a self-regulatory 'Gaian' process takes over (i.e. negative feedback, or 'virtuous circle', comes into play).

However, the need for caution about the role of plankton comes from John Woods, a leading scientist at the Natural Environment Research Council, who said that the feedback effect is positive; i.e. the ability of plankton to absorb CO_2 from the atmosphere diminishes as the warming progresses.[36] When this happens sea plants and algae emit more dimethyl sulphide. This gas is made up of tiny particles of sulphuric acid and ammonium sulphate which rise upwards to form the condensation nuclei of clouds.[37] The clouds then build up over the oceans, block out the sunlight and gradually reduce atmospheric temperatures.[38]

Indeed, so severe can this feedback be that it is believed falling levels of CO_2 alone could ultimately bring on an Ice Age. D.R. Lindstrom of the University of Illinois, and D.R. MacAyeal of the University of Chicago, experimented with a new computer model that predicted how different carbon dioxide levels might influence the extent of sea ice in the Arctic Ocean. They said the peak of the last Ice Age, about 18,000 years ago, coincided with low CO_2 levels. Their computer used data gleaned from an earlier study of ice cores drilled beneath Antarctica showing a relationship between past climates and CO_2.[39] The Milankovitch cycles were not entirely discredited, as small variations in the Earth's orbit were said to be responsible for a quite remarkably brief end to the last Ice Age, probably resulting in catastrophic flooding around the world.[40]

However, some scientists insist that something *was* going on at the very end of the Cretaceous to dramatically raise CO_2 levels. But over what timespan? Towards the end there was 'clear evidence' from carbon-13 and oxygen isotope studies of a global warming. These studies also curiously hint that 90 per cent of plankton died off at the K-T, and that this continued over 300,000 years, according to Michael Rampino and Tyler Volk of New York University.[41] This would have raised temperatures by 6°C because of the catastrophic fall in dimethyl sulphide (DMS), which, as we have seen, normally increases cloud cover. The additional carbon pumped into the atmosphere (i.e. that was no longer taken up by plankton) would make a combined rise of 10°C.

The Halsteads also imply some catastrophic end at the K-T, with sharply rising levels of CO_2, although this is not made explicit. Then they say: 'Following the sharp temperature rise there was a gradual and continuous reduction in temperature and it was probably this that was the final straw.'[42]

A Fluctuating Climate?

Leaving aside catastrophism, what can we conclude about the presumably fateful climatic events which both the uniformitarians and the catastrophists insist played a major part in the K-T extinctions? Unfortunately we cannot even arrive at a consensus on this elemental point. Only a small decline in annual temperatures may lead to a marked decline in equability, and temperatures were presumably declining throughout the Mesozoic. In the Upper Triassic, based on the discovery of red bedrocks,[43] the climate appeared to become drier and several deserts appeared. Semi-arid, subtropical scrubland had already evolved, by the early Eocene, over what is now the Sonoran Desert in California. Previously dinosaurs had roamed there in forest and savanna.[44] It was clear that seasons, and hot and occasionally arid conditions, were not infrequent.

But was the climate of the Mesozoic equable up to that point? Were there seasonal differences? The evidence, from literally hundreds of books and articles on the dinosaur environment, is, regrettably, conflicting. The Cretaceous as a whole has been described as being one of topographic and climatic smoothness. What we could tentatively conclude is that the equability and warmth of the Cretaceous came to an end somewhere in the middle of the period. We can be fairly certain that the period was warm and wet up to this point, during which time sea level was falling and many marine organisms died out.[45]

Soon, according to the general consensus, it became cooler.[46] Robert Jastrow, a distinguished American astrophysicist, said in a British newspaper interview that the 'best evidence suggests that the dinosaurs died out because of a gradual cooling of the global climate.'[47] Kenneth Hsu also says there was a global cooling during the last 100 million years, starting in the Mesozoic and ending only two million years ago,[48] and this probably affected both the distribution and the amount of precipitation.

Richard Leakey, the distinguished paleontologist, says that the beginning of the Age of Mammals 75 million years ago, was marked by 'geological stirrings and climatic instability'.[49] As a specialist concerned with mammalian primates he was interested to show how they evolved in South America as it slowly drifted westwards, with the equable temperatures favouring stable tropical/subtropical forests. Apes, he points out, evolved from pro-simian species (i.e. early monkeys), whose fossils are to be found in Africa and Asia but not in the New World. The Old World apes were subjected to a cooler climate, to

Era	*Period*	*Years ago*	*Climate*
MESOZOIC		65m	cooling
	Cretaceous		
		140m	warm/wet
		180m	
	Jurassic		warm
		195m	
	Triassic	230m	warm/dry
PALEOZOIC	Permian	280m	warm/dry
	Carboniferous	345m	tropical glacial at south pole
	Devonian	395m	warm/dry
	Silurian	435m	warm
	Ordovician	500m	cool/warm
	Cambrian	570m	cold
ARCHAEOZOIC	Algonkian	2300m	cold
	Archaean	4500m	cold

Fig.25. The climate in geological eras

broken forest and a southward drift, and to a steady drop in world temperature that began almost imperceptibly perhaps 60 million years ago, gathering momentum all the while. Daniel Axelrod says the seasonal differences were worse in the north, where they may have occurred earlier,[50] and they were less marked at low latitudes. Alan Charig says that at the end of the Cretaceous it was 'probably true' there was an increase in latitude temperature differences (i.e. according to distance from the equator) and in seasonal temperatures.[51]

We can conclude, therefore, that following the midway point there were frequent periods of wetness and aridity. Some of the world's modern deserts had their embryonic beginnings at the K-T as large drought-prone areas, as hinted at by pollen flora in California. Urey and Emiliani's 'thermometer', mentioned earlier, established also that 100 million years ago the average worldwide temperature was about 70°F — extremely warm by today's standards. It cooled to 61°F some ten million years later before rising to 70°F after another ten million years had elapsed; then it reversed again, with the record showing a slight cooling of the oceans by about 70 million years ago.[52] Winters were still mild enough for crocodilian types to survive as far north as Saskatchewan;[53] and Kenneth Hsu says deciduous tree fossils such as elm and beech hint at mild winters.

But as we get to the very end of the Maastrichtian, there seems to have been a turn-around in climate. Tony Hallam of Birmingham University believes the last few hundred-thousand years of the Cretaceous 'were marked by environmental changes more dramatic than at any other time',[54] since when it has declined steadily. Emiliani found the temperature some 30 million years ago to be about 50°F, and some 20 million years ago to be 43°F; and today it is 35°F.[55]

In the meantime the ecosystem was becoming more homogenized, and the barriers dividing terrestrial regions reduced as time passed. A lull in mountain-building activity would have resulted in vast stretches of monotonous topography, which would decrease the variety of habitats available and increase competition. Land surfaces tended to be low-lying stretches of everglade and swamp, with higher areas in between. Fewer shallow seas and shelves meant that pressure on species like the plesiosaurs increased, as they were unable to adapt to deep-sea diving (the reason why freshwater species survive in such circumstances is that retreating seas do not affect the number of lakes and rivers. Higher lakes may dry up, but lower ones are created and rivers will always drain away into the sea, even over great distances.).

Other changes in the distribution of land and sea benefited some species. Animals migrated across the Bering Straits when sea level was low enough: 70 million years ago *Protoceratops* came over to the United States with some early multi-tuberculate mammals. This less congenial environment gradually spread southwards along the varied fauna. It posed a real threat to the ecosystem, growing increasingly intricate. The new pressures seemed to impose bio-thermometric pressures on animals to favour smallness and to penalize largeness. The death knell was being sounded. Right at the end only the flat-headed forms of hadrosaur existed. Ichthyosaurs died off before the K-T event, so did *Triceratops*, but *Leptoceratops*, a contemporary of *Triceratops* from the ceratopsian stock, survived a little while longer. Gradually the *Triceratops* community, having survived a long time in close proximity with the newish type of creatures including the marsupial possums, insectivores and rodent-like multi-tuberculates, all became extinct.

Chapter 10
THE END DAYS

BEFORE we can even begin to speculate about how climatic change could have affected dinosaurs, and possibly brought about their demise, we ought to remind ourselves that dinosaurs present zoological problems that are immensely complicated. One of the most common accusations levelled at the dinosaurs, probably because it so often reflects the human condition at life's end, is that they were becoming 'senile'. In a sense this is the argument to end all arguments: they were all dying of collective old-age.

Claims of racial senility are of two sorts: the first says that certain creatures become morphologically extreme, or 'hypertelic'; the second says they become bizarrely maladaptive. The first refers to individual anatomical structures, the other relates this to their performance in the wider environment. Another word for hypertely, which has entranced evolutionary biologists for decades earlier this century, is *orthogenesis*. Theorists often point, for example, to the strangely small heads of some dinosaurs, with the implied suggestion that successful adaptation and the ability to triumph over adversity requires, over time, increasing levels of individual intelligence. Cranial capacities of dinosaurs were indeed on the small side: one type of *Brontosaurus* weighed about thirty tons, and probably had a half-pound brain. The brain of *Stegosaurus* weighed only about 70 grammes, representing only 0.0004 per cent of body weight, indicating that they were incredibly dull by present mammalian standards (although we must remember that they often had an additional spinal brain, as mentioned in Chapter 2). For instance, an elephant's brain is 0.074 per cent of its body weight, and the brain of a man — the most successful of all mammals — is 1.88 per cent.

Implied in hypertely in its second, broader sense is an inability to cope with the consequences of changed external conditions, and not just with excessive largeness or smallness of its bodily organs. Although a brain may not expand quickly enough to cope with a rapidly changing world, other parts of the body could change only too rapidly. Evolution, it was implied, occurs at an inexorable and steady rate until creatures become agents of their own extinction, with horns or teeth becoming too big for their own use. Often biologists speak of 'over specialization', meaning specialization of a kind no longer appropriate to the changed environment. One favoured example is the sabre-toothed tiger, with, in the end, curved canines so long the animals could not shut their mouths.[1]

Many academics were once sympathetic to this idea of built-in obsolescence. It did appear to some that racially senile forms developed a kind of spinescence, as in armoured dinosaurs, and perhaps toothlessness. The abnormal bony frill behind the skull of *Triceratops* could be such an example of bizarre growth, duck-billed dinosaurs had odd-looking nasal plumes, and there was the domed skull of *Pachycephalosaurus* which zoologist Cloudsley-Thompson called 'grotesque'.[2] A sizeable majority of biologists, however, rejected hypertely. More is now known about the functions of these appendages, such as the dorsal plates of *Stegosaurus* and the 'sail' of the pelycosaurs. Indeed, in spite of the changing nature of chewing techniques among dinosaurs, no fossils have been found without teeth.

The problem with the loss-of-adaptation argument is why the features that led to maladaptation afflicted one species and not another. Further, why should only the dinosaurs become senile while the other species made extinct at the K-T presumably suffered some other misfortune? Lineages exist only at any one time in history, and are of the same age. Why should some taxa or groups have very different life-spans? Surely, if one group is changing at all its last members will have changed most?

At one time those opposed to the concept of hypertely pushed the argument to the opposite extreme: they rejected totally the idea that morphological evolution runs at different 'rates'. Now the pendulum is swinging back again. Some are searching for a way to return to the idea of the 'molecular clock' popular in the 1960s: a phenomenon was said to occur whereby two related species accumulated genetic differences at a roughly steady rate. The assumption was that only a small fraction of the dna changes, unmolested by positive or negative selection, are adaptive.

The debate is now almost philosophical: if dinosaurs became extinct because of their bizarre forms, and that itself is because something is 'wrong', then one is in danger of using a circular argument if this is given as the cause of extinction. In fact the argument could be stood on its head: the most successful survivors were the most bizarre types. Take the weird forms of some ammonites, which showed spiral rather than plane coiling, or para-mammals that acquired strange reptilian-like characteristics. This happened very often at times of environmental stress; with the variations being favoured from the normal type. Aberrant forms hence may be regarded as a result of undiminished vigour and evolutionary potential in an increasingly hostile environment.[3] The complex nasal chambers of duck-billed dinosaurs were like brass instruments and could make trumpeting sounds, further implying they could communicate in a species-specific manner, and indulged in complex social behaviour.

One other example of this evolutionary potential is the growth of scaly armour which also could have acted as a substitute fur-like insulation in a cooling climate. The 'senile' forms were often the longest lived, as a result, and they were the most abundant. They co-existed well with 'youthful' forms, some of which ultimately vanished from the stratigraphic record. But not all did, and this is puzzling. Many lineages with normal skulls persisted until the very end, although significantly all of them were insulated. In fact it was the extinction of these that finally brought the dinosaur saga to a close.

One difficulty concerns the size of K-T victims and survivors, and the complicated predator/prey ratio. Dinosaurs and mammals were only marginally in competition with each other, with the dinosaurs succeeding in preventing mammals from radiating widely. This was a difficult period for the rapidly evolving ancestral mammals and paramammals,[4] and for the strange mammal-reptiles, in a class by themselves, who had reached the first stage of evolutionary development in the Triassic and were, before the K-T, on the verge of another. But the dinosaurs were in yet another class of their own, simply because, after 140 million years, they succumbed at the K-T whereas many reptilian and sub-reptilian species did not.

In fact dinosaurs were varied in size, along with their habitat. Carnivores prey on herbivores and size is often no defence against ferocious attack launched by a particularly well equipped aggressor. Despite its huge bulk the *Diplodocus* was prey to the much smaller *Allosaurus*, which had awesome prehensile feet equipped with ferocious six-inch long claws. Still, this does not necessarily imply that the herbivores were finally killed off as prey to leave the remaining carnivores to die of starvation.

The smallest dinosaurs were *Leptoceratops*, *Stegosaurus*, *Tescelosaurus* and *Dromeosaurus*, all about two to four metres long. We know that some dinosaurs also preyed on smaller reptiles like lizards: but why, in that case, did these smaller lizard-eating dinosaurs not survive, since their diet was not adversely affected? Fish-eating pterosaurs, plesiosaurs and mosasaurs vanished also. The largest known Paleocene lizard was *Peltosaurus Jepseni*, reaching a size of almost one metre. On the other hand the largest reptile surviving into the Eocene

(22 to 50 millions years ago) was *Champsaurus*, up to 1.5 metres in length. This did not survive into the Paleocene, the period immediately following the Cretaceous. Nevertheless it is curious to note nearly all of the carnivores — large and small — died out at the end.

The Problem of the Dinosaur Eggs

Dinosaur eggs have frequently featured in the extinction saga. Did the eggs, it has been asked, over time fail to hatch out in sufficient numbers? This is a question which has as its source the allegedly large number of eggs and eggshell fragments found in the Upper Cretaceous deposits of southern France. French scientists also found dinosaur eggs with pathological features in the micro-structure of the shells, appearing as if the formation was interrupted by climate extremes. Paleontologists from the University of Bonn uncovered eggshell fragments in the youngest layers of rock at a site in Provence, which were fractions of a millimetre thinner than those fragments found in older rocks.[5] Some scientists suggest that hormonal changes resulted in dinosaur eggshells becoming thinner.

Climate, weather and atmospheric conditions *do* have a bearing on the fortunes of egg-laying reptiles. Moisture control plays a crucial role in the development of eggs. Desert reptiles go to great lengths to keep incubation temperatures constant. Some prairie skink eggs are thin and flexible, unlike the hard and mineralized eggs of birds. If conditions become too wet, the eggs swell up and go mouldy, and the walls of the nest tend to collapse and become water-logged. The females incubate to ground moisture levels, partially burying the eggs. They do this more to regulate the water balance of the eggs than to warm them, because the eggs lose water to the air but gain it from the ground.[6] In wet conditions the skinks ensure the eggs get as much air as possible and lay them in nests proud of the ground.

A recent study of great tits in a Dutch forest suggests

acid rain could mean birds laying eggs with shells that are too thin. Scientists at the Institute of Ecological Research in Holland found a correlation between low calcium levels in poor soils and low calcium levels in tree foliage. Leaf-eating caterpillars, too, were calcium-poor.[7] Allaby and Lovelock link a dwindling number of dinosaur eggs with inter-species struggle. They cite the theory that the small mammals appearing during the first half of the dinosaurs' reign stole and ate all their eggs; the dinosaurs could not fight back effectively because the warm-blooded thieves were too fast and could easily dash into crevices for protection.

Bjorn Kirten dismisses the idea of egg-eating ecological disasters. He illustrates his case with the present-day Nile monitor lizard which preys on the eggs of the Nile crocodile, without actually exterminating it.[8] Egg predation goes on all the time in the modern world to no great avail. In fact reptiles lay so many eggs that most of them can be eaten without decreasing the number of adults. In any event the theory does not explain how the dinosaurs were then able to co-exist with mammals for so long a time — more than 100 million years.

Furthermore, the facts of thin dinosaur eggs need to be proven more conclusively before interesting theories are brought into the picture to 'explain' the causes of their thinning. The hundred or so eggs in the French findings are really not enough, compared with the hundreds of thousands one would expect to find. The sediments, furthermore, were very thick, indicating a very long time-span. If eggs were so sterile as to bring about extinction in one region, then the sediment would have been much, much thinner, almost non-existent. As Edwin Colbert points out, even a few hundred-thousand years will not be detectable in the geological record. Hence the extinctions appear to have been abrupt.

One feature of the 'thin eggs' thesis centres upon environmental stresses or diseases. Dietary changes as a result of newer species of vegetation emerging onto the ecological scene could result in biochemical or hormonal deficiencies.[9] This brings us to the subject of poisonous plants, long mooted as a likely dinosaur extinction theory. Ronald K. Siegel, a psychopharmacist at the University of California at Los Angeles School of Medicine, points to the introduction of harmful alkaloids or hydrolysable tannins in the early bright-coloured flowers, which began evolving toward the end of the era.[10] Adrian Desmond, a zoologist, who achieved some success by popularizing Robert Bakker's warm-blooded dinosaur thesis before Bakker published his own book, said the angiosperms could have been as toxic as strychnine or morphine, and could have produced unpleasant side-effects. They may have gained some evolutionary advantage in doing so,[11] because the dinosaurs, noticing them and having sampled them, refused to eat them. We could assume, for the sake of argument, that the dinosaurs had good palates and good colour vision, since there is no evidence to the contrary; and we do know that *Dromeosaurus* and *Deinonychus* possessed stereoscopic vision.

In truth, however, we do not know how susceptible they would have been to toxic plants, or whether they could have tolerated a moderate amount of toxins. Anthony Swain, a biochemist at the Royal Botanical Gardens, says the ferns and cycads, and many of the spiky-fronded cycadeans, employed condensed tannins but were still palatable to the dinosaurs.[12] Most of the dinosaurs had tough constitutions, and we saw in Chapter 3 how they had large digestive tracts containing detoxifying enzymes. The fossil evidence shows that many varieties of dinosaur teeth were good for crushing, and *Anatosaurus* actually ate conifer needles, as could many duckbills. Modern mammals like rabbits and horses show more tolerance to substances that Man is sensitive to, like belladonna, aconite and strychnine. But modern reptiles are able to tolerate forty times the concentration of alkaloids before discernible effects oblige them to cut down on their intake. On the other hand, modern reptiles are unable to detect relatively benign compounds such as quinine in lethal concentrations. Others have maintained it is unlikely that Cretaceous angiosperms contained enough of the alkaloids.

Some paleontologists point to the contorted pose after death of some of the long-necked dinosaurs like *Coelurosaurus*, similar to that resulting from strychnine poisoning, known as an opisthotonic condition.[13] This is caused by muscular contraction when the nervous system in death throes causes a tightening of the neck muscles and throws the head back over the body. But as Desmond reminds us, most of the dinosaur fossils were found like this, not only in the fossilized *Struthiomimus*, alive in the Cretaceous, but also in the carnivore *Compsognathus* which lived in the Jurassic before the flowering plants arrived on Earth.[14] The contorted pose could be the result of a normal drying-out of long-necked ligaments, causing them to pull back the head.

These anomalies succeed only in highlighting the long geological time interval involved: the flowering plants came into existence at the beginning of the Jurassic and the dinosaur extinctions came at the end of the Cretaceous, a time period covering at least 130 million years. There was also a relatively rapid dying off at the end, but prior to that the dinosaurs hit their peak of diversity. The gigantic herds of duck-billed hadrosaurs and the horned ceratopsians actually came into existence with the angiosperms. In fact, the ornithopod dinosaurs increased five-fold in late Cretaceous times, having enjoyed only a moderate degree of success in the late Jurassic and early Cretaceous.[15]

A curious twist to the argument suggests it was not the arrival of the flowers, by now a staple diet in the Cretaceous, but their gradual disappearance that caused the dinosaurs to starve to death. James Hopson, a dinosaur expert at the University of Chicago, says that if they ate only one plant they would be in trouble when that plant was no longer available.[16] Alan Charig says one theory of 1962 suggested that the flowers were followed by butterflies and others, whose caterpillars ate the plants. Today they are kept down by birds which did not arrive until later.

At an annual meeting of paleontologists in 1986 Robert E. Sloan of the University of Minnesota, and Keith Rigby of Nôtre Dame University, said that excavations from Montana (a region from which the vast majority of American dinosaur fossils have been found) showed that half the plants in the area died off over a period of three million years. It was this that drastically altered the dinosaurs' food supply.[17] If, incidentally, herbivorous dinosaurs had experienced a drastic reduction in their food supply, and this coincided with their final demise, it again lends weight to the argument that they were homeothermic. This is because we must multiply weight-for-weight by ten times the amount of food a homeotherm needs over that of an ectotherm.[18]

Size and Gender in a Changing Climate

One aspect of the eggshell theories leads us into a wider discussion of both ambient temperature and dinosaur gender. Two biologists at the University of Manchester, Charles Deeming and Mark Ferguson, point out that gender in reptiles is dependent on ambient temperature levels. Deeming and Ferguson noted that when the American alligator, *Alligator mississippiensis*, incubates its eggs at 30°C the hatchlings are all female, at 33°C they are all male.[19] David Crews of the University of Texas, and W. Gutzke of Memphis State University, also describe how, at 26°C, the eggs of the leopard gecko lizard all hatch as females, but at 32°C almost all become males. At intermediate temperatures both males and females develop, but the females behave in many ways as if they were males.[20]

This phenomenon is called temperature sex determination, or TSD: so gender is determined during incubation and the temperature in the nest. All lizards, crocodiles, some fish and amphibians show TSD. Although lizard and crocodile eggs hatch out males when the temperature is high, the opposite occurs with turtles where the evolutionary advantage is to lay more eggs, and so the more female eggs laid the better.

The temperature of the incubated crocodile eggs determines not only gender but growth to adulthood, as well as pigmentation and thermoregulation. Linking all these is the hypothalamus, an important control centre in the brain which co-ordinates reproductive cycles and secondary sexual characteristics, and the secretion of hormones. The advantage is that reptiles can virtually choose the gender of their offspring rather than leave it to the random chance union of sperm and egg chromosomes. The young also hatch quicker in a warm nest and are bigger too, so that adults in turn are bigger. In crocodiles this means the male is bigger when warmer, and is able better to fight off enemies and thus have a greater chance of survival. The natural sex ratios of crocodiles seem also to play an important initial role, the females having two similar X-chromosomes. A gene that directs an indeterminate embryonic sex organ to develop as a testis rather than an ovary was discovered in 1987 by David Page of the Whitehead Institute in Cambridge, Massachusetts.[21] A change of only 3°C can alter the sex of a whole clutch of eggs. This is done by the 'testis determining factor' (TDF), a gene which can bind to the dna.

But Crews and Gutzke show the TSD system to be over-rigid. A difference of a few degrees may mean the production of many physiologically female lizards which act like males and never succeed in mating; and a change in temperature of only 2°C could have skewed the sex ratio enough to affect the structure of the breeding population, with fights occurring for a mate to ensure evolutionary survival.

A recent theory decreed that the dinosaurs survived a population crash which put them at a later disadvantage; in other words, they passed through a population bottleneck and then recovered, which, according to the 40-year-old theory of Ernst Mayr of Harvard University, means the population carried on with much reduced genetic variation.[22] Scientists have long speculated as to whether large populations are more likely to make a genetic shift towards speciation than small populations. Some believed large populations would more likely undergo genetic change towards speciation because they contain more variation which has more chance of surviving changes in the environment. In the 1950s Ernst Mayr suggested that if there was too much 'genetic inertia', speciation could not take place easily because mutations would not occur frequently or fast enough — and mutations are the driving force of change.[23] Only if the population 'crashed' would it be released from its inertia.

Did this reduced genetic variation mean that the rate of dinosaur adaptation was just slower than the rate of co-evolution, leading to an unfavourable ratio in births-to-deaths over time? Could a changing climate at the K-T have produced too many of one gender? Perhaps dinosaurs lacked a skewed gender ratio they could use to exploit to their advantage. Perhaps it was warmer at the height of their evolutionary success, and declining temperatures later not only affected their gender ratios but their size. But in that case why did non-crocodilian reptiles die out? And the pterosaurs and the ammonites? Any why did other crocodilians survive? Presumably reptiles were in the temperate zones while dinosaurs were more spread out. One argument says that crocodilians survived the K-T because growth and gender were under environmental control. This allowed for rapid adjustments: TSD enabled them to scale down their size and so adapt to diminished food supply. A decrease in ambient temperature would have slowed down and lowered the 'set point' of the hypothalamus, it would have lowered the temperature at which adults preferred to live, and slowed their rates of growth.

Vertebrate paleontologist Edwin Colbert said that the large volume-to-surface ratio of alligators means that optimum temperatures could be gained by moving in and out of the Sun's rays, and he suggested that dinosaurs similarly moved in and out of tree shade. Modern reptiles, with burrowing techniques, are more adaptive to extremes of variation than dinosaurs (like varanids, turtles, lizards, snakes, etc.), but this raises a further question of why the tropical zone dinosaurs also died out. Why did not the dinosaurs compensate for a cooler climate, as do modern alligators when they seek out a warm nesting site?

The answer to this last question may depend on our understanding of animal thermal regimes. Ambient temperature is the most potent of all environmental influences, affecting every level of function — cellular activity, physiological performance, behaviour and distribution of species. Temperature has a marked influence on the biochemical and physiological reactions of all living organisms, with seasonal and diurnal rhythms affecting distribution of species populations. Extremes of temperature have been surmounted by adaptations evolved through pressure towards colonization of every available niche. For this reason the most persuasive argument for dinosaur ectothermy says that if we are looking for a cause of dinosaur evolution then it is the widespread desert conditions favouring reptilian life forms.[24]

An impressive achievement of post-thermobiology (or biometeorology) was to recognize and define the sophisticated motor activity of other reptiles, amphibians, fish spiders, etc. These are all creatures that are able to maintain a relatively constant body temperature without the aid of a 'supercharged' metabolism.[25] We now know most of the temperature extremes in which a wide range of species can survive. We know that some fish live in hot springs, but their upper limit is about 42°C, although some insects can survive in water heated to 50°C. Some bacteria can even reproduce at 70°C.

But irretrievable damage can be done to plants and animals with sub-zero temperatures. This is because ice crystals can form in their cells, and their membrane structure can be damaged.[26] In northern Labrador some fish can survive at nearly -1.8°C, which is just above freezing. The blood of most teleost fish, however, freezes at about 0.6°C. Some fish have super-cooled blood. At the other end of the thermal scale, and in the open air, Cloudsley-Thompson, testing the metabolism and heat loss of similar surface-to-volume ratios, found that at temperatures above 30°C overheating occurred even for poikilotherms. Desert plants can survive even the most harsh of conditions. Tropical plants are likely to be injured by long-term chilling in the range of 10-12°C, probably when changes in the lipid phase of their cell membranes occurs. Underground roots and tubers are more resistant to extremes than shoots, while dormant buds are more resistant than young leaves.[27] Modern plants similar to those which might have existed in the Mesozoic, thrive in regions characterized by high equability. A lower level of equability limits the northward spread of all but the most adaptable flora and fauna.[28]

Mammals often show the advantages of endothermy as the ability to maintain constant temperatures regardless of environmental gradients. On the other hand, they need to consume more energy in a cool climate, so the survival struggle is heightened. Cold-blooded vertebrates living in an environment with higher than usual temperatures experience a loading effect, necessitating a rapid or steady release of energy.[29] Conversely lower than normal temperatures mean unloading or inhibitory effects on metabolic function — creatures encounter an 'energy crisis', reducing the effectiveness of their normal functions. In other words, abnormal extremes in summer and winter would very likely threaten them with extinction.

Rapid swings of ambient temperature, however, would rarely happen like this: organisms would have plenty of time to evolve in tandem with their surroundings. For example, Cloudsley-Thompson says it is inconceivable that dinosaurs, with the gradual onset of seasons throughout the Mesozoic, did not evolve seasonal breeding cycles like other reptiles, so that hatchlings from eggs were born in summer.

Dinosaurs and Climate

The discussion in the 1930s and 1940s on metabolism of dinosaurs was conducted on the assumption that Earth was going through a warming, and that dinosaurs were largely ectothermic. In the post-war years academic opinion about both the direction in which climate was moving, and to a much lesser extent the nature of dinosaur metabolism, went through a period of revision, which has somewhat changed the parameters of the argument.

If the climate was becoming more temperate then we can now assume that it is the ectotherms which would be put most at risk, with them probably being squeezed into an ever-narrowing tropical zone. The dinosaurs would similarly suffer if they were ectothermic, and as they died out this could be proof that they were so if they were unable to migrate to the tropical zone away from the upland areas. In the light of this reasoning we can only assume that Loris Russell is confused when he says that the dinosaurs went extinct because they were endothermic, and that winter temperatures reached excessively low regimes.[30]

In fact, homeothermy or heterothermy may not have saved them from a progressively deteriorating climate. The near-mammals of the Triassic, and the Jurassic mammals such as triconodonts, symmetrodonts and pantotheres, were all probably homeotherms. Cloudsley-Thompson also draws the wrong conclusions for other reasons. He assumed (incorrectly, as we now believe) that Cretaceous temperatures were rising at the end. He therefore concentrates his argument excessively on body size and the problems associated with metabolic heat loss. He concluded that modes of life became less efficient and less economical when meteorological and geological conditions changed. He puts the main blame on the seasons, where warmer summers were coupled with increasing amounts of carbon dioxide in the air, which in turn produced greater thermal stress. However, science decrees that only if dinosaurs could move faster towards

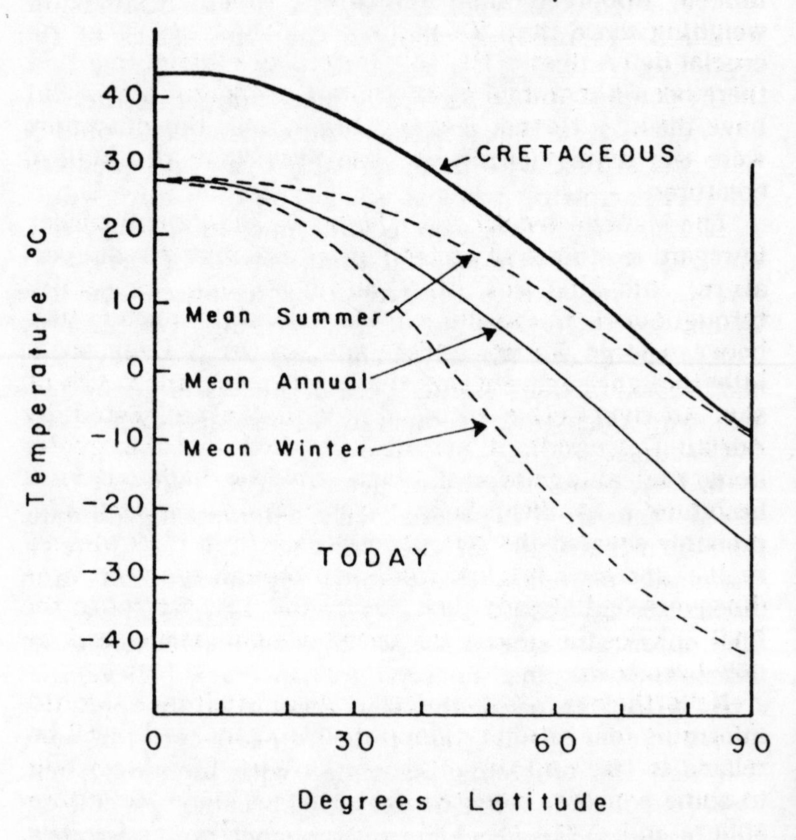

Fig.26. As early Cretaceous temperatures were considerably higher than today's, dinosaurs could have survived at latitudes nearer to the polar regions even if they had been cold-blooded. (Source: John Ostrom, AAAS Symposium, 1980)

endothermy and considerably reduce body size commensurately with the changing environment (mainly, here, climate and food supply), could they hope to keep ahead in the struggle for survival.

The curious thing about the K-T extinction, as we have seen, is that only large animals on land, and microscopic marine organisms, seemed to be affected. Indeed, Robert Bakker points out that any creature weighing more than 22 pounds vanished: size was the crucial determinant. His conclusions are intriguing: had there been a mammal of 22 pounds or more it too would have died.[31] Hence, it could be argued, the dinosaurs were the main victims because they were the biggest creatures.

The Manchester school, whome we mentioned earlier in regard to studies of reptile gender and ambient temperature, said that the dinosaurs grew big, especially throughout the Mesozoic era, because of the genetic link between large stature and rapid growth.[32] Even so, a little thought suggests that if gigantism was the result of, say, an over-active pituitary gland (as suggested by Adrian Desmond), it certainly worked to their evolutionary advantage even if a case could be made out for it becoming a handicap later. Gigantism also, incidentally, probably severed the dinosaurs at last from their 'desert-reptile' ancestral origins. And even though the very large dinosaurs had already died out by the time we get to the final ankylosaur phase, the seeds of endothermy had by now been sown.

Nevertheless a biometeorological approach would inform us that pituitary gland development could well be related to size and thermoregulation changes. According to some sources, increasing vegetation made the atmosphere more aerobic and more oxidative.[33] On the assumption that the dinosaurs had a low rate of metabolic activity, an increase in oxygen (runs the argument) would have stimulated their pituitary-adrenal system. As a result they competed more fiercely, causing a positive feedback, raising body metabolism higher and competing for more food, and so on.[34]

Whether body metabolism could be 'raised', Lamarckian-like, through a more competitive life-style is highly questionable. Tolerances had been set over 130 million years, and it would have been difficult to 'adjust' their thermoregulatory techniques as if they were machines.

A more plausible argument could be made out for a link with herbivore body-size and the availability of vegetation. We have seen already that it is a positive disadvantage for a large herbivore to be an endotherm, since they would need to eat much more than an ectotherm. Additionally, large animals suffer most acutely from competition for food supplies because they are highly specialized feeders, dependent on one or two plant species. Those smaller species that are descendants but marginally less specialized, such as birds and mammals, will have a better chance of surviving. On the other hand, great size can be of considerable advantage when food supplies start to diminish. Smaller animals need to eat daily, as anyone with a small pet knows. But large reptiles, such as the giant Galapagos tortoise, can go without food and water for up to one year.

In the last days it is possible dinosaurs became both homeothermic and smaller. Indeed the layman, awestruck by *Tyrannosaurus* and *Diplodocus* fossils mounted spectacularly in museums, often fails to appreciate the diversity of the size of the species: carnosaurs, coelurosaurs and many hadrosaurs were no larger than a full-grown turkey, and were restricted to the cooler upland areas.[35]

But smallness was still no help in the survival stakes, it seems. The hadrosaurs were already thinning out before the K-T,[36] and well-known dinosaur fossil formations were yielding less than a quarter of the number of them. Paleontologists found a gradual diminution of the genera of horned dinosaurs from sixteen (in the section before the Maastrichtian) to seven genera in the last phase. The armoured dinosaurs went from nineteen to six genera, and the duckbills from twenty-nine to seven over a period of several million years.[37]

Surprising proof of endothermy in a harsh environment comes from fossil evidence analyzed by Thomas Rich of the Museum of Victoria, and Patricia Rich of Monash University in Australia (*see* Chapter 8). It gives a new perspective to Cloudsley-Thompson's claim that at the close of the Cretaceous there were numerous dinosaurs alive and 'apparently well adapted to various habitats'.[38] These new discoveries suggest that small or medium-sized dinosaurs lived near the Earth's poles in the early Cretaceous, between 130 and 105 million years ago when average temperatures could fall to -6°C (when Antarctica, as part of *Australis*, was a great, elongated, horizontal land further to the north but still the southernmost continent).[39] Probably, for up to three months of the Antarctic winter, the dinosaurs would see no sun.

These extraordinary findings support the conclusions of researchers in the United States, who in 1987 reported dinosaurs in the Arctic.[40] And it also validates the claims of Robert Bakker who said that fossils laid down in cooler continents have since drifted to warmer zones, thus 'proving' they were endothermic. Thomas and Patricia Rich say the dinosaurs may have coped by migrating, but many were permanent residents of these colder regions of the Earth. Herbivores would have had a supply of evergreen herbaceous plants in winter and deciduous plants in summer. They also identified new plant-eating species of ornithopods called hypsilophodonts, which lived there for tens of millions of years. The Riches also found three or four species of meat-eating theropods (probably predators of the hypsilophodonts, and pterosaurs) as well as freshwater plesiosaurs.[41]

Most scientists agree there can be no single obvious explanation for the final extinction of the ceratopsians and the ankylosaurs. Most of the physiological theories of extinction we have discounted already, such as senility, and gender imbalance. We have tentatively come to the conclusion that climatic factors were involved, and that dinosaurs had a variety of thermoregulatory techniques.

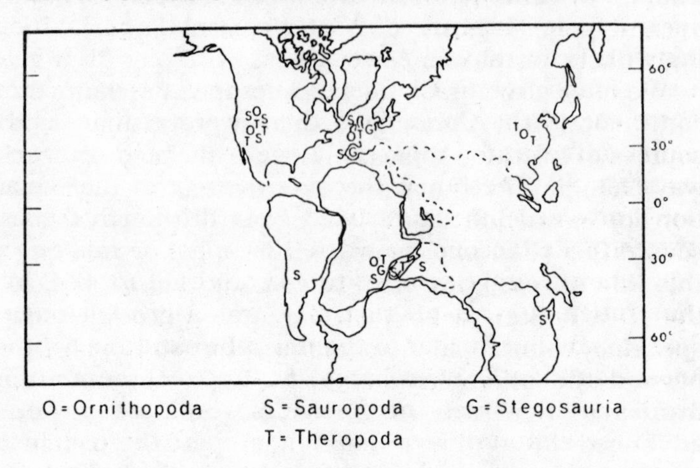

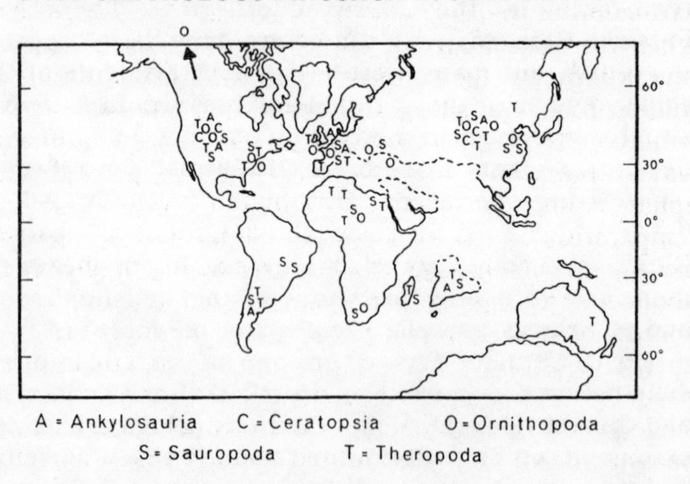

Figs 27 & 28. Note the widespread distribution of dinosaur fossils.

Many of them, as we have seen, may have been homeothermic (not quite the same as being warm-blooded; *see* Chapter 4), although in the light of fossil discoveries near ancient polar regions even endothermy seems increasingly likely for the smaller species.

We have also agreed that, over millions of years due to continental drift, climate was cooling, and that possibly sources of nutrition were diminishing — or in some other way changing for the worse. In other words our explanation for extinction lies in ecological or geophysical reasons rather than physiological, zoological or genetic, and this line of reasoning had, of course, been advanced by the 'Death Star' school of thought, although, strictly speaking, catastrophic explanations for extinction lie well beyond the parameters of any of the aforementioned frames of reference. It is, in a sense, an argument advanced *in extremis*; one that, somewhat like the senility one, succeeds in foreclosing all further debate — the final *coup de grâce*.

Let us, then, in the dying pages of this book and before we look more reflectively at the subject of dinosaur extinction in the Epilogue, say that the 'final straw' came when the missile struck Earth. Firstly the missile, about one kilometre in diameter and weighing about 100 million tons, travelling at a speed of 25 km per second, would generate devastating kinetic energy and shock waves that would arise from an explosion in the atmosphere equivalent to 100,000 million tons of TNT.[42] The temperature of the atmosphere in the first few minutes, as the missile passed through it, would have reached about 200°C, and scorching winds would have reached speeds of 400 km per hour. Shards of blistering hot energy would have burst across the surface, melding rock particles and triggering volcanic eruptions. The next stage would be for the atoms of gas in the atmosphere to be rearranged; i.e. nitrogen and oxygen would have been blended to become oxides of nitrogen. These could soon become nitric acids which produce nitrates in the soil, especially in conjunction with heavy rains.

Using the recent Kilauea eruption as a model, Terrence M. Gerlach of the Sandia National Laboratory in Albuquerque, estimated the Deccan Traps injected up to 30 trillion tons of carbon dioxide, six trillion tons of sulphur, and 60 billion tons of chlorine-type substances into the lower atmosphere, to linger in the air for a few hundred years.[43]

It is here that, already, a period of revision has taken place in the 'impact extinction' thesis. One of the major arguments advanced by Luis Alvarez and his colleagues initially was that a massive dust veil would have been lofted into the skies as a result of the destruction of the bolide as it finally made violent contact with the Earth's surface. The theory was that the subsequent darkening of the skies as the dust veil circled the Earth for something like three months or more, would effectively halt photosynthesis on both the land (firstly) and then the sea, and would ultimately destroy plant life, thus causing the ultimate starvation to any herbivores surviving the impact. This theory is now considered to be wrong, since the dust cloud would likely fall back to Earth within hours or at most days. The main destructive element, especially for vegetation, would be a combination of toxicity of the surface, and widespread forest fires, and for aquatic creatures to be killed off due to the toxicity of the seas. A widespread greenhouse effect due in part to the raised level of CO_2 arising from the global fires and partly from the death of phytoplankton in the seas, would have meant, overall, that many creatures would either have been starved, destroyed on impact, or would have died of thermal stress or been poisoned. Indeed, a recent twist to the K-T extinction theory was advanced by Thomas Wdowiak, an astrophysicist at the University of Alabama, who suggested that nickel, from the vaporized asteroid, poisoned the plants on which the dinosaurs fed.[44]

EPILOGUE

THE READER may, at the end of this book, be experiencing a vague sense of disappointment. The 'Fate' in the title turns out to be what some might say is a truism and others a facile abstract generalisation — that the universe is in a sense running down and that all species eventually either die off without issue or simply transmute into yet other similar, perhaps more advanced, species. Yet I have resisted giving a non-committal run-down of all the known theories of dinosaur extinction: a few of the more well-known theories appear in the text, but only within a developed framework of explanation. I believe there are approximately one hundred distinct 'explanations' of dinosaur extinction, most of which I have long since forgotten. Stephen Jay Gould once wrote that the catalogue of proposals would 'fill a Manhattan phone book'; Drs Halstead and Charig, in a lively public debate in London in August, 1983 cantered briefly through fifty or more theories, some published in books and journals but most simply mooted in academic common-rooms, and not a few others in public bars.

It is not that the vast majority of extinction theories are absurd, since many phenomena and processes on Earth are absurd as well as being cruel and pointless (including, incidentally, the 'impact extinction' event), but that they cannot patently be true if the evolutionary story is understood in detail. No single theory can simply be plucked out of the air and applied to fit just one species, or taxa or genera, at any one period in earth history. Any theory must explain not only why the dinosaurs died, but why other co-existing species, like many types of reptile, the turtles, the crocodilians, birds and mammals, miraculously survived; and why, at the same time, a host of other species, especially marine organisms, all died (and why, for that matter, did the land extinctions lag behind the marine extinctions by some half a million years?). Any extinction theory must explain not only why virtually none of the larger genera survived, but also why many of the smaller ones perished. In other words, the explanation must take full account of the selective nature of the Great Dying, and it must be scientifically plausible.

Many theories make sense only when applied to individual species, or, a fine paradox, to too many. Dinosaur theorists are in conflict as to whether the dinosaurs were out-competed by the succeeding generation of mammals. One refutation of this thesis was the fact that the dinosaurs were still diversifying into new orders while the mammals did not begin to follow suit until the Great Extinction. Nevertheless the role played by mammals in the mystery is intriguing, and crucial to our understanding of events. They survived even though it took some while before they began to proliferate and diversify. There were thriving bands of marsupials (whose young are raised in pouches, like kangaroos) and placentals (whose young are born fully formed).

The mammals may have survived because, being smaller, they can evolve at a faster rate than very large animals like most of the dinosaurs. The bigger the animal, the longer is its life-cycle from infancy to adulthood. By contrast a small creature can pass through many generations while a large form is undergoing only one. So a small organism can evolve much further in a given time, whereas a larger form cannot adjust itself so rapidly to change. And yet the Great Dying finished off not only the larger and more specialized dinosaurs, like *Tyrannosaurus*, *Triceratops* and the ankylosaurs, but also the small, unspecialized ones like the coelurosaurs and hypsilophodonts.

Even in 1953 George Gaylord Simpson, the eminent American paleontologist, was writing about the 'old, long and futile argument' that tried to focus either on biological or external factors. A way forward, he said, was to assume that extinction was *usually* caused by loss of adaptation to a changing environment. He took it as axiomatic that the environment changed continually in some degree, even minute by minute, requiring commensurate adaptive powers on the part of the species. Extinction, therefore, is a function of a maladaptive relationship between the two phenomena.[1] Certainly, he said, environmental change by itself could not be the cause of extinction.

Species evolve because the planet evolves. They are part of the stream of life, having no independent essence of their own, simply paving the way for other species. As Stephen Gould points out, life is not a tale of progress, but instead a story of intricate branching off and wandering, with the fortunate taxa surviving and, all being well, adapting.[2] Gould rejects the notion of species-death as a form of failure. Extinctions are as much a pattern of life as evolution. They are the 'enabling force' of the biosphere — the ultimate fate of all lineages. Species simply fail to adapt speedily enough to changes in climate, or to competition.[3]

One of the laws of nature decrees that any decisive movement away from the equilibrium state leads to irreversible changes: even if some kind of balance is later restored. The consumption and redistribution of matter and energy will proceed along an altered state because the world is constantly changing. And it changes precisely because of the on-going process of redistribution.

This observation shows up the superficiality of many extinction theories. Take the 'cosmic' explanations, upon which a great number of dinosaur books in recent years have concentrated in an unashamedly partisan way.*

* Many authors, such as Kenneth Hsu, Richard Muller and David Raup, were of course closely involved in the formulation of the 'Death Star' theory, and have simply described their role in the development of certain aspects. At the same time, they have provided a keen insight into how science is 'done'.

Cosmic catastrophe would certainly bring about a decisive and possibly terminal movement away from the equilibrium; but the idea that a giant bolide from space landed on Earth, or a supernovae explosion took place, probably fails not only because the evidence upon which the case rests can be interpreted in another way, but also because there is no *irrefutable* evidence that the event actually occurred some 65 million years ago. True, the meteorite theory might just be able, given time, to withstand this criticism, but it is unlikely the supernovae theory could. The weakness of the latter theory is that the lake-bottom dwellers would have been little affected, when clearly they were the most severely affected.

A different sort of criticism can be levelled at the 'cataract in the eyes' thesis, which comes under the 'dinosaur disease' or 'faulty gene' category of explanation. Here the problem is that the (remaining) dinosaurs *all* became extinct, and it is unlikely that diseases would affect an entire species in this way. The species may be reduced in numbers, but when the survivors become thinly scattered the ability of the disease to continue spreading is limited. Then the parasite and host adjust to each other, with the parasites dying out each time a host animal dies. Similarly, the most disease-resistant strains of the species increase at the expense of the more prone members, until an immune species evolves.[4] Many other theories concerning dinosaur gender, metabolism, the strength of dinosaur eggshells and so on, are dammed because the extraordinary longevity of the dinosaur species makes one wonder why they did not succumb much, much sooner. The same criticism applies to the theory that the dinosaurs became progressively constipated as the more fibrous and oily gymnosperms were replaced by the angiosperms. This theory fails because angiosperms were around for literally tens of millions of years before the dinosaurs became extinct.

Incidentally, very few animal species are excessively specialized when it comes to diet. Ant-eaters would perish if ants and termites disappeared; pandas would likewise if bamboo shoots disappeared; the koala eats only the leaves of certain kinds of eucalyptus: but most herbivores are partly also insectivores and eat a variety of vegetation, and most carnivores will eat any animal smaller than themselves, and some small mammals like rodents can eat a wide range of insects, nuts, fruit and roots.

Changing habitat theories are similarly flawed because habitats are always changing. We are here, however, edging closer to the truth because extinction is clearly coterminous somehow with geophysical change. Stephen Jay Gould reminds us that falling sea-levels accompanied virtually every known mass extinction.[5] Climatic change theories, in particular, stand out as the one explanation that withstands criticism well. Even these theories, however, could be seen to be inadequate since the long time-span involved would be as prolonged as the time needed for a species to evolve, and hence adapt accordingly. Nevertheless we know that climates can change over tens or hundreds of thousands of years, while species take millions of years to evolve. Further, we know that climate is a product of continental drift, and that evolution seems to go at full throttle at the maximum rate of drift. Could we not, then, speculate that there might have been a faster rate of change at the K-T? It is at this stage that the more catastrophic theories begin to grow in explanatory power and plausibility: they become, in a sense, the ultimate explanations, tidying up all the loose theoretical ends. They say, in effect, that the rapid speeding up of all the factors that would put a species at a disadvantage — insufficient in themselves to cause extinctions — meant that many of the species could no longer adapt quickly enough to ensure survival.

The catastrophists, furthermore, succeed in highlighting the doomsday forces that predominate in shaping life on Earth. The dinosaurs had the seeds of death sown deep within them from the very dawn of the solar system. The Earth, as well as showing evolutionary design, shows paradoxically the signs of chaos, as recent discussion of 'chaos theory' has shown. Global turbulence is incurred in both cosmic and terrestrial events and a variety of ecological catastrophes, some major, some minor, some occurring with disturbing regularity and others with great infrequency. This implies that the catastrophic and uniformitarian perspectives are not so distant, and in any event the two views were always too simplistic. Darwin himself knew that the endless waves of evolution, though always subject to the laws of Nature, at times receded to become a stagnant millpond, and at other times roared ahead to become a Niagara.

Footnotes

Chapter 1

1. Fred Hoyle trilogy, *Lifecloud* and *Diseases from Space*, Sphere Books, London, 1979; and *Evolution from Space*, Paladin Books, London, 1983.
2. John Macvey, *Where Will We Go When The Sun Dies?*, Stein & Day, New York, 1983.
3. *Ibid*, p.72.
4. Ian Ridpath, *Signs of Life*, Penguin, London, 1977.
5. John Gribbin, *Genesis*, Dent, London, 1980, p.86.
6. *New Scientist*, 4 March 1989.
7. Macvey, *op.cit.*
8. Lloyd Motz, *The Universe — Its Beginning and End*, Abacus, London, 1977.
9. Michael Wolfson and John Dormond, *The Origin of the Solar System*, Ellis Horwood, Chichester, 1989.
10. *Discover* magazine (US), May 1987; see also Barrow and Tipler, *The Cosmic Anthropic Principle*, Clarendon, Oxford, 1986.
11. John Gribbin and Martin Rees, *The Stuff of the Universe*, Heinemann, London, 1990.
12. *Discover* magazine (US), *op.cit.*
13. James R. Beerbower, *Search for the Past*, Prentice-Hall, New Jersey, 1968, p.145.
14. John Gribbin and Stephen Plagemann, *The Jupiter Effect*, Macmillan, London, 1974.
15. Preston Cloud, *Cosmos, Earth and Man*, Yale University Press, New Haven, 1978.
16. R.L. Rosenburg *et al.* in *Nature* magazine, Vol.250, 1974.
17. Lyn Margulis and Dorian Sagan, *Microcosmos*, Allen & Unwin, London, 1987.
18. James Lovelock, *Gaia*, Oxford University Press, Oxford, 1979.
19. *New Scientist*, 17 March 1988.
20. Antony Milne, *Our Drowning World*, Prism Press, Bridport, 1989; see also Fred Pearce, *Turning Up The Heat*, The Bodley Head, London, 1989.
21. *Science 85* magazine (US), October 1985.
22. David Schwartzman and Tyler Volk in *Nature* magazine, Vol.340, 15 August 1989.

23. Cloud, *op.cit.*, p.114.
24. Macvey, *op.cit.*, Chapter 1.
25. Cloud, *op.cit.*, p.109.
26. *New Scientist*, 6 February 1986.
27. *Nature* magazine, Vol.342, 9 November 1989.
28. *The Daily Telegraph*, 19 February 1990.
29. Barrow and Tipler, *op.cit.*, p.511.
30. Fred Hoyle, *Lifecloud*, Sphere Books, London, 1979.
31. Margulis and Sagan, *op.cit.*
32. Cited in Fred Warshofsky, *Doomsday*, Abacus, London, 1979, p.150.
33. Peter Russell, *The Awakening Earth*, Routledge, London, 1982, p.30.
34. James A. Lake in *Nature* magazine, 31 January 1988.
35. Fred Hoyle, *Lifecloud*, *op.cit.*
36. Isaac Asimov, *A Choice of Catastrophes*, Hutchinson Press, London, 1979, p.209.
37. *The Times*, 31 October 1990.
38. Robert Jastrow, *Until the Sun Dies*, Fontana, London, 1979, p.58.
39. Michael Allaby and James Lovelock, *The Greening of Mars*, André Deutsch, London, 1984.
40. Magnus Pyke and Patrick Moore, *Everyman's Facts and Feats*, Dent, London, 1981.
41. Peter Cook and John Shergold in *The Guardian*, 12 July 1984.
42. *Proceedings of the National Academy of Sciences* (US), May 1989.

Chapter 2

1. J.R. Beerbower, *Search for the Past*, Prentice-Hall, New Jersey, 1968, p.46.
2. Jeremy Cherfas in *The Independent*, 28 September 1987.
3. Beerbower, *op.cit.*, p.47.
4. Richard Dawkins, *The Blind Watchmaker*, Longmans, London, 1986.
5. Stephen Jay Gould, *Hens' Teeth and Horses' Toes*, Penguin, London, 1984, p.255.
6. Stephen Jay Gould, *The Flamingo's Smile*, Penguin, London, 1987, p.444.
7. Gordon Rattray Taylor, *The Great Evolution Mystery*, Secker & Warburg, London, 1982, p.18.
8. *Ibid*, p.51.
9. *The Independent*, 27 October 1986.
10. Jane Burton and Dougal Dixon, *Age of Dinosaurs*, Sphere

Books, London, 1984, p.16.
11. Lyn Margulis and Dorian Sagan, *Microcosmos*, Allen & Unwin, London, 1987.
12. Beerbower, *op.cit.*, p.140.
13. Stephen Jay Gould, *Wonderful Life*, Century Radius, London, 1990.
14. Taylor, *op.cit.*
15. David Dineley, *Earth's Voyage Through Time*, Paladin, London, 1975.
16. Paul Davies, *The Cosmic Blueprint*, Heinemann, London, 1987, p.108.
17. Dineley, *op.cit.*, p.75.
18. *Ibid*, p.211.
19. *Ibid*, p.206.
20. *Science* magazine (US), Vol.246, 1989, pp.1953-1959.
21. *New Scientist*, 19 November 1988.
22. Taylor, *op.cit.*, p.58.
23. *Ibid*, p.66.
24. T.F. Gaskell and Martin Morris, *World Climate*, Thames & Hudson, London, 1979, p.109.
25. Sprague De Camp, *Day of the Dinosaur*, Bonanza, New York, 1985, p.89.

Chapter 3

1. Michael Allaby and James Lovelock, *The Great Extinction*, Secker & Warburg, London, 1983, p.14.
2. L.B. Halstead, *The Evolution and Ecology of the Dinosaurs*, Peter Lowe, London, 1975.
3. Allaby and Lovelock, *op.cit.*, p.14.
4. A. Charig, *A New Look at Dinosaurs*, Heinemann, London, 1979, p.22.
5. *Ibid*, p.13.
6. David Norman, *Illustrated Encyclopedia of Dinosaurs*, Salamander Books, London, 1985.
7. Charig, *op.cit.*, p.18.
8. John C. McLoughlin, *Archosaurs*, Penguin, London, 1979, p.29.
9. Charig, *op.cit.*, p.18.
10. Robert Bakker, *The Dinosaur Heresies*, Longman, London, 1986, p.209.
11. Charig, *op.cit.*, p.18.
12. Norman, *op.cit.*, p.11.
13. *The Times*, 5 January 1990.
14. Alan Charig and Brenda Horsfield, *Before the Ark*, BBC Books, London, 1975.

15. *Ibid*
16. Charig, *op.cit.*, p.11.
17. Richard Dawkins, *The Blind Watchmaker*, Longmans, London, 1986, p.266.
18. James R. Spotila, ed., *AAAS Selected Symposium 28*, Roger Thomas & Everett C. Olson, 1980, p.234.
19. Jeremy Campbell, *Winston Churchill's Afternoon Nap*, Aurum Press, London, 1986, p.262.
20. *The Times*, 14 September 1984.
21. *Nature* magazine, Vol.341, November 1989.
22. Stephen Jay Gould, *Ever Since Darwin*, Burnett Books, London, 1978, p.175.
23. J.B.S. Haldane, *On Being the Right Size*, Oxford University Press, Oxford, 1985.
24. Gould, *op.cit.*, p.173.
25. Haldane, *op.cit.*, p.1.
26. *The Daily Telegraph*, 31 December 1988.
27. Norman, *op.cit.*
28. *Discover* magazine (US), September 1989.
29. McNeill Alexander, *The Times Higher Education Supplement*, 7 December 1990.
30. Halstead, *op.cit.*, p.21.
31. Tony Thulborn, *Dinosaur Tracks*, Chapman & Hall, London, 1990.
32. McLoughlin, *op.cit.*, p.21.
33. *Ibid*, p.48.
34. *Science Digest* (US), October 1985, p.12.
35. Halstead and Halstead, *Dinosaurs*, Blandford Press, London, 1981.
36. Bakker, *op.cit.*, p.215.
37. *Ibid*, p.218.
38. John Ostrom, *AAS Symposium*, *op.cit.*, p.34.
39. Bakker, *op.cit.*, p.128.
40. *Discover* magazine (US), October 1990.
41. *The Guardian*, 12 October 1990.
42. Charig and Horsfield, *op.cit.*
43. Charig, *op.cit.*, p.98.
44. Halstead and Halstead, *op.cit.*, p.22.
45. David Norman and David Weishampel in *New Scientist*, 7 May 1987.
46. Halstead, *op.cit.*, p.28.
47. McLoughlin, *op.cit.*, p.52.
48. Bakker, *op.cit.*, p.184.
49. Charig, *op.cit.*, p.100.
50. Halstead and Halstead, *op.cit.*, p.28.
51. Bakker, *op.cit.*, p.180.
52. *Ibid*, p.173.

53. *Ibid*, p.196.
54. *Ibid*, p.169.
55. Sprague De Camp, *The Day of the Dinosaur*, Bonanza, New York, 1985, p.125.
56. Ostrom, *op.cit.*, p.171.
57. Norman and Weishampel, *op.cit.*
58. Charig and Horsfield, *op.cit.*
59. Norman, *op.cit.*
60. Sprague De Camp, *op.cit.*, p.189.
61. Kenneth Hsu, *The Great Dying*, Harcourt Brace-Jovanovich, New York, 1987, p.51.
62. *The Observer*, 30 July 1989.
63. Hsu, *op.cit.*, p.106.
64. Hsu, *op.cit.*, Ballantine, New York, p/b 1988, p.52.
65. Duncan Lunan, *Man and the Stars*, Souvenir Press, London, 1974, p.54.
66. Hsu, *op.cit.*, Ballantine, New York, p/b 1988, p.52.

Chapter 4

1. Charig, Bellais and Cox, eds, *Morphology and Biology of Reptiles*, Linnean Society, No.3, 1976.
2. Beverly Halstead in *New Scientist*, 6 August 1987, p.59.
3. *Ibid.*
4. John Ostrom, ed., *AAAS Selected Symposium 28*, Roger Thomas & Everett C. Olson, 1980, p.15.
5. Philip Regal, *AAAS Selected Symposium 28*, *ibid*, p.178.
6. *Ibid*, p.171.
7. *Ibid*, p.181.
8. E.J.W. Barrington, *Environmental Biology*, Edward Arnold, London, 1980, p.142.
9. *Ibid*, p.148.
10. *Ibid*, p.151.
11. John Noble Wilford, *The Riddle of the Dinosaur*, Alfred A. Knopf, New York, 1986, p.57.
12. *Ibid*, p.58.
13. *Ibid*, p.60.
14. *Journal of Paleontology*, Vol.39, No.3, May 1965.
15. Robert Bakker, *The Dinosaur Heresies*, Longman, London, 1986, p.354.
16. Ostrom, *op.cit.*, p.41.
17. J.L. Cloudsley-Thompson, *Why the Dinosaurs Became Extinct*, Meadowfield, London, 1978.
18. *New Scientist*, 6 August 1987, p.59.
19. Ostrom, *op.cit.*, p.53.
20. Ian Anderson in *New Scientist*, 24 September 1987, p.25.

21. Ostrom, *op.cit.*, p.43.
22. *Ibid*, p.40.
23. Cloudsley-Thompson, *op.cit.*
24. Charig, Bellais and Cox, *op.cit.*
25. John C. McLoughlin, *Archosaurs*, Penguin, London, 1979, p.30.
26. Ostrom, *op.cit.*, p.52.
27. Peter Wheeler in *New Scientist*, 12 May 1988, p.62.
28. Halstead and Halstead, *Dinosaurs*, Blandford Press, Poole, 1981, p.145.
29. *Ibid*, p.23.
30. Ostrom, *op.cit.*, p.39.
31. Halstead and Halstead, *op.cit.*, p.24.
32. McLoughlin, *op.cit.*, p.30.
33. Wheeler, *op.cit.*, p.63.
34. Sprague De Camp, *Day of the Dinosaur*, Bonanza, New York, 1985, p.97.
35. McLoughlin, *op.cit.*, p.26.
36. *The Guardian*, 12 October 1990.
37. Ostrom, *op.cit.*, p.36.
38. Regal, *op.cit.*, p.180.
39. *Nature* magazine, Vol.340, 1989, p.138.
40. Cloudsley-Thompson, *op.cit.*
41. Stephen Jay Gould, *The Flamingo's Smile*, Pelican, 1987, p.431.
42. Beverly Halstead, *The Evolution and Ecology of Dinosaurs*, Peter Lowe, London, 1975.
43. James C. Spotila, *AAAS Symposium*, *op.cit.*, p.234.
44. Regal, *op.cit.*, p.181.
45. Kenneth Hsu, *The Great Dying*, Harcourt Brace-Jovanovich, New York, 1987, p.109.
46. Nicholas Hotton, *AAAS Symposium*, *op.cit.*, p.37.
47. Sprague De Camp, *op.cit.*, p.123.
48. Ostrom, *op.cit.*, p.31.
49. *Ibid*, p.27.
50. McLoughlin, *op.cit.*, p.68.
51. Beverly Halstead in *New Scientist*, 6 August 1987.
52. *AAAS Symposium*, *op.cit.*
53. David Norman, *Illustrated Encyclopedia of Dinosaurs*, Salamander Books, London, 1985.
54. E. Carol Roth, *AAAS Symposium*, *op.cit.*, p.21.
55. Spotila, *ibid.*
56. Sprague De Camp, *op.cit.*, p.62.
57. *Discover* magazine (US), January 1991.
58. Spotila, *op.cit.*, p.234.
59. *Discover* magazine (US), March 1989.

Footnotes

Chapter 5

1. David Raup, *The Nemesis Affair*, W.W. Norton & Co., New York, 1986.
2. *New Scientist*, 1 December 1990.
3. Beverly Halstead, *The Evolution and Ecology of Dinosaurs*, Peter Lowe, London, 1975.
4. Van Valen, *Paleobiology*, Vol.10, 1984, pp.121-37.
5. Richard Muller, *Nemesis*, Weidenfeld & Nicolson, New York, 1988, p.13; see also *Journal of Geological Education*, 1, 986, Vol.34, p.90.
6. Michael Allaby and James Lovelock, *The Great Extinction*, Secker & Warburg, London, 1983.
7. Daniel Jablonski, *The Last Extinction*, eds. Les Kaufmann & Kenneth Mallory, MIT Press, Cambridge, Massachusetts, 1986, p.56.
8. *Ibid*, p.57.
9. John Man, *The Day of the Dinosaur*, Bison Books, 1978.
10. Muller, *op.cit.*, p.13.
11. John Taylor, *Science and the Supernatural*, Temple Smith, London, 1980, p.40.
12. Cedric Woods in *New Scientist*, 10 April 1988.
13. Jim Shirley, cited in Antony Milne, *Earth's Changing Climate*, Prism Press, Bridport, 1989, pp.36, 37.
14. Flood and Lockwood in *The Nature of Time*, ed. Roger Penrose, Basil Blackwell Ltd, Oxford, 1986, p.56.

Chapter 6

1. Isaac Asimov, *A Choice of Catastrophes*, Hutchinson Press, London, 1979, p.212.
2. Bob Whitten and Sheo Prasod, *Ozone in the Free Atmosphere*, Van Nostrand Reinhold, New York, 1985.
3. Claud C. Albritton Jr, *Catastrophic Episodes in Earth History*, Chapman & Hall, London, 1989, p.121.
4. Whitten and Prasod, *op.cit.*, p.272.
5. *Ibid*, p.275.
6. *Ibid*, p.198.
7. Environmental Protection Agency, *UV radiation and the environment*, Washington, 1986.
8. Whitten and Prasod, *op.cit.*, p.276.
9. *Ibid*, p.277.
10. National Academy of Sciences (US), 'Causes and Effects of Changes in Stratospheric Ozone', 1984, p.215.
11. Whitten and Prasod, *op.cit.*, p.275.
12. *New Scientist*, 29 July 1989, p.48.

13. *Nature* magazine, 8 May 1989.
14. David Jablonski, *The Last Extinction*, eds Les Kaufmann & Kenneth Mallory, MIT Press, Cambridge, Mass., 1986.
15. John Maddox in *The Times*, 31 August 1989.
16. *Nature* magazine, *op.cit.*
17. *Ibid.*
18. *New Scientist*, 29 July 1989.
19. Antony Milne, *Earth's Changing Climate*, Prism Press, Bridport, 1989, p.23.
20. *Ibid*, p.49.
21. Ian Crain, quoted *ibid*, p.51.
22. Albritton, *op.cit.*, p.123.
23. Frank Close, *End: Cosmic Catastrophe*, Simon & Schuster, New York, 1988, Chapter 9.
24. *The Times*, 30 November 1984.
25. Boyle and Ardrill, *The Greenhouse Effect*, New English Library, London, 1989, p.82.
26. David Croft, *The Last Dinosaurs*, Elmwood Books, 1982.
27. John C. McLoughlin, *Archosaurs*, Penguin, London, 1979.
28. *New Scientist*, 22 July 1989.
29. H.E. Hinton, *Illusion in Nature and Art*, Gerald Duckworth, London, 1973.
30. *The Times*, 2 August 1988.

Chapter 7

1. Stephen K. Donovan, ed., *Mass Extinctions: Processes and Evidence*, Belhaven Press, London, 1989.
2. Michael Allaby and James Lovelock, *The Great Extinction*, Secker & Warburg, London, 1983, p.32.
3. *Science* magazine (US), Vol.227, 8 March 1985.
4. Antony Milne, *Earth's Changing Climate*, Prism Press, Bridport, 1989, p.106.
5. Kenneth Hsu, *The Great Dying*, Harcourt Brace-Jovanovich, New York, 1987, p.166.
6. Richard Muller, *Nemesis*, Weidenfeld & Nicolson, New York, 1988, p.44
7. *Ibid*, p.66
8. *Ibid*, p.83.
9. *Ibid.*
10. *Scientific American*, October 1990.
11. *New Scientist*, 18 February 1989.
12. Muller, *op.cit.*, p.82.
13. Hsu, *op.cit.*, p.170.
14. *Nature* magazine, Vol.285, 1980.
15. Hsu, *op.cit.*, p.175.

16. *Ibid*, p.178.
17. *Nature* magazine, Vol.339, 8 June 1989.
18. Milne, *op.cit.*, p.133.
19. John Noble Wilford, *The Riddle of the Dinosaur*, Faber & Faber, London, 1986, p.251.
20. *New Scientist*, 12 November 1988.
21. Isaac Asimov, *A Choice of Catastrophes*, Hutchinson Press, London, 1979, p.142.
22. *Ibid*, p.138.
23. *The Times*, 21 April 1989.
24. Muller, *op.cit.*, p.227.
25. Asimov, *op.cit.*, p.133.
26. Frank Close, *End: Cosmic Catastrophe*, Simon & Schuster, New York, 1988, p.37.
27. Hsu, *op.cit.*, p.148.
28. *Ibid*, p.147.
29. Muller, *op.cit.*, p.107.
30. Hsu, *op.cit.*, p.154.
31. Close, *op.cit.*, p.38.
32. Wilford, *op.cit.*, p.42.
33. *Nature* magazine, 18 May 1987.
34. Donald Goldsmith, *Nemesis*, Walker & Co., New York, 1985, p.119.
35. Whitmire and Matese in *Nature* magazine, Vol.313, 1985.
36. Muller, *op.cit.*, p.177.
37. *The Times*, 1 July 1987.
38. *The Daily Telegraph*, 4 December 1989.
39. Muller, *op.cit.*, p.107.
40. John W. Macvey, *Where Will We Go When the Sun Dies?*, Stein & Day, New York, 1983, p.100.
41. Muller, *op.cit.*, p.255.
42. Goldsmith, *op.cit.*, p.136.
43. Claud C. Albritton Jr, *Catastrophic Episodes in Earth History*, Chapman & Hall, London, 1989, p.146.
44. *Science* magazine (US), 16 March 1984.
45. Michael Rampino in *New Scientist*, 18 March 1989, p.55.
46. *New Scientist*, 12 November 1988.
47. *New Scientist*, 18 March 1989.
48. *New Scientist*, 10 December 1988.
49. *New Scientist*, 12 November 1988.
50. *New Scientist*, 24 November 1990.
51. *New Scientist*, 17 November 1990.
52. *Science* magazine (US), Vol.244, July 1989, pp.1565-67.
53. *The Times*, 7 October 1988.
54. Hsu, *op.cit.*, p.221.
55. T.F. Malone and J.G. Roederer, eds. *Global Change*, Cambridge University Press, Cambridge, 1985.

Chapter 8

1. *National Geographic*, October 1990.
2. Emanuel Velikovsky, *World in Collision*, Abacus, London, 1972.
3. Carl Sagan, *Broca's Brain*, Hodder & Stoughton, London, 1979.
4. Antony Milne, *Our Drowning World*, Prism Press, Bridport, 1988, p.103.
5. R. Fairbridge and J. Shirley, *Solar Physics*, December 1987.
6. *Geology* magazine, Vol.17, 30 September 1989, p.661.
7. John Gribbin, *This Shaking Earth*, Sidgwick & Jackson, London, 1978.
8. Goesta Wollin, cited in Antony Milne, *Earth's Changing Climate*, Prism Press, Bridport, 1989, p.58.
9. Michael Shallis, *On Time*, Burnett Books, London, 1982.
10. Isaac Asimov, *New Guide to Science*, Penguin, 1987, p.365.
11. *Ibid*, p.367.
12. *Discover* magazine (US), November 1987.
13. *Omni* magazine (US), July 1987.
14. *Ibid*.
15. *Discover* magazine (US), November 1987.
16. Preston Cloud, *Cosmos, Earth & Man*, Yale University Press, New Haven, 1978, p.98.
17. Paul Andersen and Deborah Cadbury, *Imagined Worlds*, BBC Books, London, 1985, p.130; see also Cloud, *op.cit.*, p.78.
18. *Omni* magazine (US), July 1987.
19. Rob Butler in *New Scientist*, 21 October 1989.
20. Kenneth Hsu, *The Great Dying*, Harcourt Brace-Jovanovich, New York, 1987, p.53.
21. *Ibid*, p.54.
22. *Ibid*, p.65.
23. Edwin H. Colbert, *Dinosaurs*, Hutchinson Press, London, 1962.
24. Robert Bakker, *The Dinosaur Heresies*, Longman, London, 1986.
25. *Nature* magazine, Vol.329, 24 September 1987.
26. Hsu, *op.cit.*, p.91.
27. Bakker, *op.cit.*, p.438.
28. *Science* magazine (US), 12 May 1989.
29. F.A. Middlemass *et al.*, *Faunal Provinces in Space and Time*, Seel House Press, 1971, p.112.
30. David Jablonski, *The Last Extinction*, eds. Kaufmann and Mallory, MIT Press, Cambridge, Massachusetts, 1986, p.57.
31. Bakker, *op.cit.*, p.439.

32. David Dineley, *Earth's Voyage Through Time*, Paladin, London, 1975, p.307.
33. Alan Charig and Brenda Horsfield, *Before the Ark*, BBC Books, London, 1975.
34. *The Times*, 29 December 1988.

Chapter 9

1. Kenneth Hsu, *The Great Dying*, Harcourt Brace-Jovanovich, New York, 1987, p.58.
2. *Discover* magazine (US), May 1989.
3. *Geography* magazine, October 1989.
4. Claud C. Albritton Jr, *Catastrophic Episodes in Earth History*, Chapman & Hall, London, 1989, p.156.
5. J.C. Crowall, *Climate in Earth History: Studies in Geophysics*, National Academy Press, Washington, 1982.
6. Hsu, *op.cit.*, p.52.
7. Albritton, *op.cit.*, p.157.
8. Stephen Jay Gould, *Ever Since Darwin*, Burnett Books, London, 1978, p.137.
9. Thomas J. Schopf, *Journal of Geology*, March 1974.
10. *Ibid.*
11. Steven Stanley, *Extinction*, Scientific American Library, 1987, p.116.
12. Charles R. Pelligroni and Jesse A. Stoff, *Darwin's Universe*, Van Nostrand Reinhold, New York, 1983.
13. Albritton, *op.cit.*, p.163.
14. Steven Stanley, *Scientific American* (US), 1984, p.350.
15. *New Scientist*, 8 September 1990.
16. *Science* magazine (US), Vol.206, 12 December 1979.
17. *Nature* magazine, Vol.219, 1981, p.650.
18. Isaac Asimov, *New Guide to Science*, Penguin, London, 1987, p.188.
19. Erick J. Barron *et al.* in *Science* magazine (US), Vol.212, 1 May 1981.
20. Stefan Gartner *et al.* in *Science* magazine (US), Vol.206, 14 December 1979.
21. *Nature* magazine, 31 August 1989.
22. Kai Petersen, *Prehistoric Life on Earth*, Methuen, London, 1978, p.110.
23. Adrian Desmond, *The Hot-Blooded Dinosaurs*, Blond & Briggs, London, 1975.
24. Leo J. Hickey in *Nature* magazine, Vol.292, 1981, pp.529-31.
25. Van Valen and R.E. Sloane, *Evolution Theory* magazine, 1977, 2, 37, p.44.

26. Hsu, *op.cit.*, p.106.
27. *Biological Journal of the Linnean Society*. Vol.36, May 1989, pp.227-49.
28. Van Valen and Sloane, *op.cit.*, p.44.
29. *New Scientist*, 7 July 1990, p.27.
30. *Biological Journal of the Linnean Society, op.cit.*
31. *New Scientist*, 5 November 1987.
32. *New Scientist*, 24 August 1990.
33. D.M. McLean in *Science* magazine (US), Vol.210, 1978, p.410.
34. *Nature* magazine, Vol.340, 24 August 1989.
35. Hsu, *op.cit.*, p.107.
36. *The Times*, 4 September 1989.
37. Fred Pearce, *Turning Up the Heat*, The Bodley Head, London, 1989, p.90.
38. *Nature* magazine, 16 April 1987.
39. *Science* magazine (US), Vol.245, 11 August 1989.
40. Antony Milne, *Floodshock*, Alan Sutton, Gloucester, 1986.
41. *New Scientist*, 17 March 1988.
42. L.B. and J. Halstead, *Dinosaurs*, Blandford Press, London, 1981.
43. Ronald Pearson, *Climate and Evolution*, Academic Press, London, 1978, p.137.
44. Daniel I. Axelrod and Harry P. Bailey, *Evolution* magazine, Vol.22, 1968.
45. Tony Hallam in *Nature* magazine, Vol.251, 18 October 1974.
46. Barron Thompson *et al.* in *Science* magazine (US), Vol.175, 1981, and Vol.212, 1982; see also Van Valen and Sloane, *op.cit.*; also J.A. Wolfe in *American Scientist*, No.66, 1978, p.694; also V.A. Krassilov in *Paleontology*, No.21, 1978, p.893.
47. *The Daily Telegraph*, 21 March 1988.
48. Hsu, *op.cit.*, p.105.
49. Richard E. Leakey and Roger Lewin, *Origins*, Macdonald Futura, London, 1982, p.36.
50. Hickey, *op.cit.*
51. A. Charig, *A New Look at Dinosaurs*, Heinemann, London, 1979, p.151.
52. Charles R. Pelligroni and Jesse A. Stoff, *Darwin's Universe*, Van Nostrand Reinhold, New York, 1983.
53. John Man, *The Day of the Dinosaur*, Bison Books, 1978.
54. Hallam, *op.cit.*
55. Asimov, *op.cit.*

Footnotes

Chapter 10

1. *New Scientist*, 10 February 1990.
2. Cloudsley-Thompson, *Why the Dinosaurs Became Extinct*, Meadowfield, 1978.
3. John Man, *The Day of the Dinosaur*, Bison Books, 1978.
4. Kai Petersen, *Prehistoric Life on Earth*, Methuen, London, 1978.
5. Edwin H. Colbert, *Dinosaurs*, Hutchinson Press, London, 1962.
6. *Zoological Journal of the Linnean Society*, March 1989.
7. P.J. Drent and J.W. Woldendorp in *Nature* magazine, Vol.339, 8 June 1989.
8. Bjorn Kirten, *Age of Dinosaurs*, World University Library, 1968.
9. Fred Warshofsky, *Doomsday: The Science of Catastrophe*, Abacus, London, 1977.
10. *Time* magazine, 6 May 1985.
11. Adrian Desmond, *The Hot-Blooded Dinosaurs*, Blond & Briggs, London, 1975.
12. T. Swain *et al.*, eds. *Linnean Society Symposium No.3*, 1976.
13. Desmond, *op.cit.*, p.188.
14. *Ibid.*
15. *Ibid*, p.187.
16. *Time* magazine, *op.cit.*
17. *The Times*, 12 November 1986.
18. Michael Allaby and James Lovelock, *The Great Extinction*, Secker & Warburg, London 1983, p.124.
19. *New Scientist*, 25 March 1989.
20. *Nature* magazine, 26 April 1988.
21. *Cell*, Vol.51, 1987, pp.1104-09.
22. *American Naturalist*, Vol.134, 1990, p.668.
23. *New Scientist*, 10 February 1990.
24. David Norman, *Illustrated Encyclopedia of Dinosaurs*, Salamander Books, London, 1985.
25. Andrew R. Cossins and Ken Bowler, *Temperature and Biology of Animals*, Chapman & Hall, London, 1988.
26. E.J.W. Barrington, *Environmental Biology*, Edward Arnold, London, 1980.
27. Charles R. Pelligroni and Jesse A. Stoff, *Darwin's Universe*, Van Nostrand Reinhold, New York, 1983, p.166.
28. *Journal of Paleontology*, Vol.39, No.3, 1965.
29. Robert Bakker, quoted in John Man *op.cit.*
30. *New Scientist*, 25 March 1989.
31. A. Schatz, *Proceedings of the Pennsylvania Academy of Sciences*, 32:26-36.

32. *Ibid.*
33. Lorris Russell in *Journal of Paleontology*, Vol.42, No.1, 1967.
34. Colbert, *op.cit.*, p.193.
35. Man, *op.cit.*
36. Cloudsley-Thompson, *op.cit.*
37. *New Scientist*, 8 April 1989.
38. Elizabeth Brouwers in *Science* magazine (US), Vol.237, 1987.
39. *New Scientist*, 8 April 1989.
40. Antony Milne, *Earth's Changing Climate*, Prism Press, Bridport, 1989, p.142.
41. *Scientific American*, October 1990.
42. *New Scientist*, 19 May 1990.

Epilogue

1. Gaylord Simpson, *Major Features of Evolution*, Columbia University Press, New York, 1953.
2. Stephen Jay Gould in *Natural History* magazine, August 1982.
3. Stephen Jay Gould, *Ever Since Darwin*, Burnett Books, London, 1978.
4. Sprague De Camp, *Day of the Dinosaurs*, Bonanza, New York, 1985, p.201. 5. Stephen Jay Gould, *Hens' Teeth and Horses' Toes*, Penguin, London, 1984, p.323.

Classifications

Dinosaurs mentioned in this book, classified by orders, sub-orders and infra-orders.

SAURISCHIANS (herbivores and carnivores)

Theropods

Allosaurus
Ceratosaurus
Coelurosaur
Compsognathus
(coelurosaur)
Carnosaur
Carnotaur
Deinonychus
Dilophosaurus
Megalosaurus
(carnosaur)
Tyrannosaurus
Stenonychosaurus
Struthiomimus
Dromeosaurus

Sauropods

Brachiosaurus
Brontosaurus
(apatosaurus)
Diplodocus
Supersaurus
Prosauropod
Anchisaur
Plateosaur
Saltosuchus

ORNITHISCHIANS (herbivores)

Ornithopods

Anatosaurus
Camptosaurus
Hadrosaurus
Hypsilophodont
Iguanodon
Leosothosaurus
Ornithosuchus
Proceneosaurus
(hadrosaur)
Saurolophus
(hadrosaur)
Tescelosaurus
Trachodon

Pachycephalosaurs

Pachycephalosaurus

Ceratopsians

Tetraceratops
Triceratops
Protoceratops
Leptoceratops
Psittacosaurus
Styracosaurus
Torosaurus

Stegosaurs

Stegosaurus

Ankylosaurs

Ankylosaurus
Nodosaur

Glossary

abiotic process: theory that life on Earth and the biosphere is controlled by the physical Earth itself, in particular the amount of carbon in the soil and atmosphere.
algae: tiny aquatic plants.
Algonkian period: between 670 million and 600 million years ago.
ammonite: extinct coiled shellfish abundant in the Mesozoic era.
amniote egg: embryo of creature that develops within an amniote membrane or shell.
amphibian: creature able to live both in water and on land.
anapsid: a reptile group characterised by having no skull openings behind the eye socket.
angiosperm: flowering ('covered seed') plant.
Anthropic Principle: theory that intelligent life selects out its own universe.
archosaurs: a grouping of higher vertebrates based on certain shared skeletal features; includes dinosaurs and modern crocodilians.
arthropods: animals with jointed legs; e.g. crabs, insects and spiders.
asteroid: similar to meteorite, but considerably larger, some reaching several hundreds of miles in diameter.
bacteria: microscopically small organisms.
belemnite: fossil shell of an extinct cephalopod.
biometeorology: the study of creatures as they function within different climatic zones.
biped(al): an animal that habitually stands or walks on its hind legs.
bolide: a solid object from space reduced to plasma by the violent heat of entry through the atmosphere.
brachiopod: shelled sea creature resembling the clam and oyster.

Cambrian period: the most ancient of the Paleozoic time zones in which animal life first appeared.
carnosaur: bipedal carnivore of the Saurischian order.
catastrophism: theory that both evolutionary events on Earth and the extinction of species are brought about by catastrophic and violent means.
cell: simplest living organism of which larger organisms are built.
Champsaurus: early aquatic lizard-like reptile.
climate: long-term, statistical change in weather patterns over considerable lengths of geologic time, measured in hundreds of years.
co-evolution: changes in species or larger groups as the wider environment changes.
cold-welding: process whereby specks of dust aggregate into larger lumps, and eventually planets.
comet: object from space with solid nuclei of icy rocks trailing a long gaseous tail.
conifer: cone-bearing tree, such as fir, pine and yew.
continental drift: theory that parts of the Earth's surface have drifted toward and away from each other throughout geologic time.
Copernican Principle: theory that Earth is no more privileged than any other part of the universe.
cosmic rays: fast-moving subatomic particles such as alphas and protons that arrive at the surface of the Earth from outer space.
cotylosaur: early amphibious creature common in the Carboniferous and Permian periods.
Cretaceous period: the last period in the Mesozoic era, lasting 70 million years; some 65-100 million years ago.
crossopterygii: alleged to be one of the forerunners of the lungfish.
cycad: palm-like plant possessing a single stem and a crown of fern-like leaves, flourishing in the Triassic and Jurassic.
cynodont: advanced type of mammal-like reptile of the Triassic period.
Deccan Traps: famous site of ancient lava-flows in India.

Devonian period: period in which fish emerged onto land, 395-345 million years ago.
diapsids: a reptile group having a pair of openings immediately behind the eye socket; includes dinosaurs and crocodilians.
dimetrodon early mammal-like reptile, of the Pelycosaur ('sail reptile') order.
dipnoan: modern lungfish, still extant in Africa and Australia.
ecology: the study of the relationship between living things and the environment.
ectothermic: 'cold-blooded'; reliant on external heat sources for raising internal temperature of creatures.
endothermic: 'warm-blooded'; able to generate heat internally to raise body temperature.
entropy: law of science that says things pass from a low state of disorder to a higher state, as time passes.
Eocene era: epoch in the Tertiary period, 55-37 million years ago.
ET (or *extraterrestrial*): belief that major events on Earth, such as extinctions, are caused by missiles or bolides landing from outer space.
euparkeria: an archosaurian reptile, said to be an early forerunner of the dinosaurs, common in the Triassic.
eustasy: the rise and fall of sea levels as ice-caps wax and wane over time.
evolution: a gradual change in the characteristics of species by natural selection.
extinction: the final death of an entire species.
family: a grouping of creatures in a similar genera.
feedback: where additional inputs into a system either regulate it (negative feedback), or exacerbate it (positive feedback).
femur: upper leg or thigh bone.
fenestra: window-like opening in the skull of an archosaurian or reptilian creature.
flora: plants and vegetation.
foraminifera: single-celled planktonic organisms with a hard shell.

fossil: petrified remains of animals or plants in rock-forming sediments.
Gaian hypothesis: theory that biological Earth processes and the ecosystem are self-regulating. (From Gaia, Greek goddess of the Earth.)
gastroliths: stones used to grind up food in an animal's digestive system.
gene: unit of inherited material in the cell nucleus having a constant effect on the development of a creature.
genus: a grouping to which several species belong.
geophysics: a modern continuation of geology, including a knowledge of plate tectonics, internal Earth pressures and dynamics to explain climatic change, volcanism and earthquakes.
gigantism: tendency for species to grow excessively large; thought to be a cause for extinction in some.
Gondwana: the 'southern continents' in the Triassic, comprising Antarctica, South America and Australia.
graviportal: slow-moving, lumbering.
greenhouse effect: theory that solid planets are made warmer at the surface by additional inputs of carbon dioxide into the atmosphere.
gymnosperms: 'naked seed' plants (for example, the non-flowering), including grasses and cereals.
herbivore: plant-eating animal.
homeothermy: a form of regulation in creatures, in which temperatures remain roughly steady. Thought to be a 'halfway house' between ectothermy and endothermy.
hypertely: theory that some creatures become racially senile with anti-evolutionary and bizarrely shaped features.
hypsilophodontid: herbivorous ornithopod of the Ornithischian order.
ichthyosaur: streamlined marine reptile of the Mesozoic era, with fish-shaped body.
ichthyostegid: early amphibian with weak stumpy legs.
ilium: one of the bones of the pelvis.
invertebrates: soft-bodied creatures, without backbone, such as worms, sponges, insects.

iridium: heavy metal grains found in meteorites, Earth's core and in stratified rocks.
ischium: one of the pelvic bones, like the ilium, which helps geologists determine whether a dinosaur is an ornithischian or saurischian.
isotope: chemically identical atoms differing only in their weight and stability.
Jurassic era; middle era of the Mesozoic period, 190-140 million years ago, during which dinosaurs were at their most widespread.
K-T: period demarcating the end of the Mesozoic (the end of the Cretaceous) with the beginning of the Tertiary (the Paleocene), beyond which no dinosaur fossils can be found.
labyrinthodont: squat, four-footed armoured amphibian, common in the Triassic and Permian.
Lamarckism: mechanism of evolution advanced by French naturalist J.B. de Monet de Lamarck (1744-1829), who believed that characteristics (like calluses) acquired during an animal's lifetime could be passed on to subsequent generations.
Laurasia: the 'northern continents' of the Triassic, comprising North America, Europe and Asia.
Maastrichtian era the last few million years of the Cretaceous.
mammal: animal that suckles its young.
mantle: region of the Earth's interior between the outer crust and the core.
Mesozoic period: one of the main geologic periods, 65-270 million years ago; referred to as the Age of Reptiles.
metabolism: chemical processes in the body that break down food to provide energy, varying in efficiency between ectotherms and endotherms.
meteorite: small rocky object from space, often heavily blended with metals.
micro-organism: the simplest living matter, unicellular; such as bacteria.
mosasaurs: group of reptiles that appeared at the end of the Cretaceous.

natural selection: the adaptation and survival of species to perpetuate their kind; i.e. the environment 'selects out' the fittest organisms.
nodosaur: armoured dinosaur of the Ankylosaur sub-order.
omnivore: an animal with a diet of both plant and animal food.
Ordovician period: 465-475 million years ago, between the Cambrian and the Silurian.
ornithischia: one of two major groups of dinosaurs, comprising ornithopods, stegosaurs, ankylosaurs and ceratopsians.
ozone: form of oxygen with three atoms of oxygen instead of two; the 'ozone layer' is several miles up in the strato-sphere.
pachycephalosaur: bone-headed dinosaur of the Ornithi-schian order.
paleontology: the interpretation of past animal habits and characteristics as revealed by the fossil record.
Paleozoic era: period in which early life forms originated, 600-225 million years before the Mesozoic.
Pangaea: the enormous supercontinent formed in late Permian times when all the continents collided.
Panspermia: the theory that precursors of organic life drifted to Earth from outer space.
paramammal: earlier forerunner of the mammal still evolving from the reptilian stage.
parapsid: a reptile group characterised by a single open-ing in the skull, high up on the top.
pelycosaur: mammal-like reptiles ('sail backs') of the Carboniferous and Permian periods.
periodicity: theory that events on Earth occur at regular or cyclical intervals.
Permian era: the last period in the Paleozoic, 280-215 million years ago, giving way to the Triassic.
photo... (prefix): biological activity carried out with the aid of the Sun, or as a reaction to the Sun's rays.
placodont: heavily armoured turtle-like reptiles of the Triassic.

planetesimal: small, planet-like body in space.
plate tectonics: the large moving plates which make up the Earth's crust.
plesiosaur: marine reptile of the Mesozoic era, with large flippers.
predator: animal that hunts prey for food.
proterosuchian: crocodilian thecodont of the late Permian period.
pterosaur: flying reptile of the Mesozoic era, distantly related to the dinosaurs.
pubis: one of the pelvic bones that differs in ornithischians and saurischians.
quadruped: an animal that walks on all fours.
rhynchosaur: large, pig-like reptile related to the lizard, from the late Triassic.
saurischian: one of the two major groups of dinosaurs, comprising all theropods and sauropods, the latter being both carnivorous and herbivorous.
secondary plate: bone forming the roof of the mouth, to separate it from the nasal cavity.
seymouriamorph: important transitional animal halfway between an amphibian and a reptile.
species: a group of animals that look the same and can interbreed.
stratigraphy: the study of the layered (stratified) patterns in rock.
supernova: a star that has reached the end of its life and explodes.
synapsid: a reptile group characterised by a single opening in the skull behind the eye socket.
T Tauri stars: glowing clouds of dense cosmic dust, the precursors of real stars.
Tertiary period: first period of the Mesozoic, 225-200 million years ago, when dinosaurs first appeared.
Tethys: an ancient sea that separated Laurasia from Gondwana; a remnant of this sea is the Mediterranean.
tetrapod: vertebrate with four limbs.
thecodont: Triassic reptile with teeth in sockets, including pseudosuchians, aetosaurs and phytosaurs.

therapsid: mammal-like reptile of the late Permian and early Triassic.
thermoregulation: techniques of regulating the internal heat of mammals, reptiles and amphibians.
theropod: predatory saurischian dinosaur, mostly bipedal.
Triassic period: 345-195 million years ago.
ultraviolet: solar radiation just above the visible spectrum, operating at a frequency of 10^{16} cycles per second.
vertebrates: animals with backbones, such as fish, amphibians, reptiles, birds and mammals.

Select Bibliography

Albritton, Claud C. Jr. *Catastrophic Episodes in Earth History*, Chapman & Hall, London, 1989.

Bakker, Robert. *The Dinosaur Heresies*, Longmans, London, 1986.

Charig, Alan. *A New Look at Dinosaurs*, Heinemann, London, 1979.

Cloudsley-Thompson, J. *Why the Dinosaurs Became Extinct*, Meadowfield, 1978.

Dawkins, Richard. *The Blind Watchmaker*, Longmans, London, 1986.

Gould, Stephen Jay. *Wonderful Life*, Century Radius, London, 1990.

Halstead, Beverly. *The Evolution and Ecology of Dinosaurs*, Peter Lowe, London, 1975.

Lovelock, James. *Gaia*, Oxford University Press, Oxford, 1979.

Margulis, Lyn and Dorian Sagan. *Microcosmos*, Allen & Unwin, London, 1987.

Muller, Richard. *Nemesis*, Weidenfeld & Nicholson, New York, 1988.

Norman, David. *Illustrated Encyclopedia of Dinosaurs*, Salamander Books, London, 1985.

Raup, David. *The Nemesis Affair*, W.W. Norton & Co., New York, 1986.

Stanley, Steven. *Extinction*, Scientific American Library, 1987.

Taylor, Gordon Rattray. *The Great Evolution Mystery*, Secker & Warburg, London, 1982.

Wolfson, Michael and John Dormond. *The Origin of the Solar System*, Ellis Horwood, Chichester, 1989.

Index

Page numbers in italic refer to illustrations.

Index

Index

Index